MISSION TO MARS

MISSION TO MARS

Plans and Concepts for
the First Manned Landing

James E. Oberg

STACKPOLE BOOKS

Published by
STACKPOLE BOOKS
Cameron and Kelker Streets
P. O. Box 1831
Harrisburg, PA   17105

Opinions and evaluations in this book are those of the author and cannot be
construed to represent the views of the author's employers, McDonnell Douglas
or the National Aeronautics and Space Administration.

Printed in the U.S.A.

Library of Congress Cataloging in Publication Data

Oberg, James E., 1944–
  Mission to Mars.
  Bibliography: p.
  Includes index.
  1. Space flight to Mars.   I. Title.
TL799.M3023   629.45'53   82-5689
ISBN 0-8117-0432-7   AACR2

To the Mars Underground—
who believe it will happen.

"Here it is
The ragged coast—the coast that no one knows.
How far the lands march inland?
                                No one knows.
Is there a Northwest Passage to the East?
Is there Cathay beyond? Can Englishmen
Live there and plant there and breed there?
                                No one knows.
And yet, I know this much. It must be tried.
My one man's life hath seen this England grow
Into a giant from a stripling boy
Who fenced about him with a wooden sword
And prattled of his grandsire's wars . . .
—The long, the ruinous wars that sucked us dry,
. . . nightmare, endless wars,
. . . Then we turned seaward. Then the trumpets blew.
And, suddenly, after the bloodshot night
. . . and the gropings in the dark,
There were new men, new ships and a new world.
And yet, how did we dare, how did we dare!
How did we dare to send our sailors out
Beyond all maps?

Stephen Vincent Benét, *Western Star* (1943),
Imagining the tale of Thomas Smyth, an
Elizabethan merchant/explorer

Contents

Foreword

TWO HOURS BEFORE Neil Armstrong and Edwin Aldrin landed on the moon, Sir Francis Chichester—who had sailed his fifty-three-foot boat *Gypsy* solo around the world—remarked, "They've got something I envy them enormously for. And that is that one of them is going to pull off one of the greatest firsts in history."

The first manned lunar landing is, however, only one of three great "firsts"—manned landing on the moon, manned landing on another planet, and manned landing on a planet around another star.

Men have landed on the moon several times. Future activity there will be devoted to thorough exploration leading to utilization. We can anticipate this within the next twenty-five years because the moon possesses valuable resources.

But what of the second great "first," the first manned landing on another planet?

Planners have long known that, of all the inner planets, Mars is the most likely target. Mercury orbits too close to the sun and is extremely hot. Unmanned Soviet space probes have landed on

Venus and reported an extremely hostile environment. Mars, on the other hand, can be tackled with a modest improvement and foreseeable development in technology. It has a cold, thin, dry carbon dioxide atmosphere.

Because of these characteristics, "Ruddy Mars" has beckoned the imaginative and drawn the attention of astronomers for a long time. There have been many books, magazine articles, technical papers, and project proposals written about manned Mars landing activities. As the years have gone by and space technology has matured, each successive work has shown a growing maturity, too. This book by James Oberg presents the most recent findings and current thinking.

Travel to the moon was once the epitome of impossibility. Now it can be done, albeit expensively. Soon it will be economically feasible.

A manned Mars landing used to be something totally beyond discussion, so patently impossible did it seem. Now it is possible but very expensive, the same evaluation given to a manned lunar landing in 1950. By the end of the century, it will not only be possible but necessary because the human race will be on the verge of using the Solar System for economic purposes. We will have to go to Mars to find out what is there that has value. And within fifty years, travel to Mars will be as commonplace as transatlantic travel in 1982.

Beyond Mars and the planets lie the stars. The journey to Mars will not only provide many of the answers to the problems of going to the stars, it will also perform a most important function: getting our eggs out of one planetary basket.

The dreams and aspirations of the space pioneers have not been concentrated solely on the moon. During the early 1930s upon his morning arrival in the basement shops of the Moscow Group for the Study of Reactive Motion, Soviet space pioneer Fridrkh Arturovich Tsander would shout to his colleagues, "Onward to Mars!"

The long-range planners of the Great Space Diaspora echo Tsander's cry today.

G. Harry Stine

Acknowledgments

MUCH OF THE material in this book is based on information distributed at the landmark Case for Mars colloquium held April 30 through May 2, 1981, on the campus of the University of Colorado, Boulder. Frequently I will simply refer to such and such as being discussed by someone "at Boulder," or "during the 1981 Boulder Mars colloquium." The conference, sponsored by the Mars Study Group (primarily a group of graduate students associated with the Laboratory for Atmospheric and Space Physics) and other space advocate groups such as the Planetary Society, was a tremendous success in drawing together numerous disparate, isolated activities from NASA, industry, university, and private groups and individuals into a mutually supportive alliance which is still gathering strength.

Much new material was gathered in my own research, from special reports prepared over the past fifteen years, some on serious man-to-Mars plans (Geoffrey Cannetti at North American Aviation, now Rockwell International, or Morris Jenkins at the NASA Johnson Space Center in Houston), some post-Viking strategies (Donald

11

Bogard et al., in Houston, plus other groups centered around James Cutts, Hal Masursky, and Ronald Greeley). These reports are very limited in distribution at best, and in a few cases I worked with the only known surviving copies.

Special thanks is due to the Case for Mars enthusiasts, Carol Stoker, Christopher McKay, Penelope Boston, Tom Meyer, Steven Welch, and others, and to everyone who came and helped make the meeting a success, particularly Humboldt Mandel, Dick Johnson, Lou Friedman, Jim Cutts, Ben Clark, Conway Snyder, Daniel Woodard, Stan Kent, Bassett Maguire, Phil Quattrone, Buzz Aldrin, Jim French, Leonard David, "BJ" Bluth, Fred Singer, and others. There were analysts and enthusiasts who could not come but sent their results: Hal Masursky, Peter Vajk, John Stevenson (who died in 1980), Bob Parkinson, G. Harry Stine, Don Cox, Arden Albee, Everett Gibson, Krafft Ehricke, and others. Special technical help was provided by specialists such as Joe Kosmo, Morris Jenkins, Carl Seyfert, Uel Clanton, and others.

From NASA public affairs, enthusiastic help above and beyond the call of duty came from Don Bane (JPL), Lisa Vazquez, Mike Gentry and Terry White (NASA Houston), and Pete Waller (Ames Research Center); historians Ed Ezell (Houston) and Lee Saegesser (headquarters) also dug up a lot of bones. Special thanks once again to Fran Waranius, Ron Weber, Stephen Tellier, and other staffers at the Lunar and Planetary Institute in Houston.

Artists such as Paul Hudson, David Hardy, David Egge, Adolf Schaller, and Patrick Rawlings helped convert many ideas into concrete images.

For reviewing the manuscript and correcting obvious errors, I wish to thank Kathryn Sullivan, William Fisher, Keith Holden, and Karl Henize. Any errors which remain are wholly the fault of the author.

And thanks especially to my wife Cooky and son Gregory who shared me with this project, and provided crucial encouragement and understanding when I seemed to be on some other planet.

Introduction

THIS BOOK IS about manned exploration and settlement of Mars. It is not a book about the planet Mars itself, although the planet is the focus of the plans for getting there and its characteristics are continually involved in designing the mission. It is not a book about space colonies, although Mars itself will be shown to be a potential site for human settlement. Nor does the book deal with other peripheral issues of space flight such as warfare, or extraterrestrial intelligence. Here you will find the theme of sending people to Mars, beginning about twenty years from now, and doing it as part of an expanding effort leading to the permanent human presence on another world.

Here is the punch line right up front: a manned (and by "manned" I will mean both men and women as space travellers) expedition to Mars is technologically feasible as soon as we decide to do it, and can be justified on scientific, economic, social, and political grounds. Furthermore, it can be carried out by the end of this century with an equivalent expenditure of about half of what it took in the 1960s to send Apollo astronauts to the moon.

The progression of space technology developed by Earth's spacefaring nation-states in the 1980s and 1990s will have a logical and inevitable consequent, manned interplanetary flight, and the only choice facing the United States is whether to participate in this process (and, by virtue of U.S. technology and experience, do so in a leading role) or to step aside and allow perhaps still undefined alliances of other national-corporate-military-ideological forces to take the lead in this and other aspects of Earth's future.

The default of the United States in this arena of human activity will not prevent the establishment of a multiplanetary civilization for twenty-first century humankind, but merely (at worst) delay it a few decades, while insuring that such a future will not necessarily reflect the characteristics and values of contemporary North American civilization, nor benefit the descendants of today's Americans.

This is *not* a hard blueprint for any particular "first man-to-Mars" expedition. Rather, the book is a discussion of many factors, both obvious and subtle, which will affect the design of such a future mission strategy. They constitute the body of assumptions, guidelines, and constraints which any future man-to-Mars plan will have to accommodate.

As such, the book reviews the past quarter century of evolution of proposed manned mission strategies and concepts, picking and choosing those ideas which are likely to continue to be relevant at the turn of this century. At the same time, it discusses the kinds of interim unmanned probes to Mars needed to fill in the gaps of knowledge so as to insure the success of the manned expeditions described.

Next, the transportation question is tackled: first, with a view of defining the actual space routes to Mars and back, and then by describing the kinds of hardware needed for propulsion. The most favorable early missions can make a surprising shortcut past Venus, and remain at Mars for a few months at most, while soon-to-follow expeditions could stay for more than a year with even less propulsion requirements than the shorter trips. No "wild card" propulsion systems are needed: a combination of chemical stages and aerodynamic braking can accomplish a manned Mars mission with a vehicle whose assembly in Earth parking orbit needs no more than four or five launchings of a heavy booster derived from equipment already proved on Space Shuttle missions.

Human factors will be challenging but not impossible. Life-support systems will be improvements of those expected to be already operational aboard permanent space stations in the 1990s. Medical problems are far less of a concern after recent Soviet six-month manned space missions, but space radiation (particularly from solar flares) is definitely a factor to be worried about. The outbound leg of the trip, far from being boring and uneventful, will be crammed full of training tasks and other educational activities, along with housekeeping, exercise, and some uniquely valuable scientific work.

The scientific goals of Mars exploration will involve the origin of the solar system and the evolution of planets and life upon them. Comparative planetology has already revolutionized the scientific notions about our home planet, and that revolution can be expected to continue and expand, with valuable harvests of earth-side applications. Meanwhile, Mars promises to make available unique records of recent climate-change factors in the solar system, information vital to the anticipation of potentially disastrous influences on Earth's own climate. To answer these questions, specific landing sites and exploratory profiles have already been mapped out.

The small and too easily overlooked moonlets, Phobos and Deimos, offer surprisingly attractive options to support man-to-Mars. They are easier to reach and return from than is Mars itself, and offer both secure shelter and accessible raw materials for the on-site production of air, water, and even rocket propellant. With a fuel depot on one of the moons, the economy of the Earth-Mars-Earth route is vastly improved. Also, a single-stage, liquid-fuelled, state-of-the-art martian reusable rocket vehicle could shuttle supplies, personnel, and raw materials from the moonlet to the martian surface and back with remarkable ease and economy.

Activities on the martian surface may center around transportation capabilities, which can be made surprisingly potent. A dual jeep system with portable shelters can allow astronauts to range hundreds of miles from their landing point. Supported by small cache drops from an orbiting mother ship, the Mars explorers can roam up to one thousand miles away from base on weeks-long traverses. Back at the base, one of the major early activities would be to set up pilot plants for the extraction of usable life-support consumables from the environment; later, fuel processing and food gardens would follow.

Highly credible cost analysis techniques show that the unique expense of a man-to-Mars program would be far less than that of the Apollo man-on-the-Moon program of the 1960s, in terms of real dollars, federal budget percentage, or per capita cost. This is not fiscal finagling: the rationale for this pleasant surprise is that most technological capabilities required for a Mars expedition would already have been developed and funded for other concurrent space activities, whereas Apollo had to buy the entire spectrum of space hardware all at once. This implies that a single major country, or a consortium of small countries, could indeed afford such a program. The main contender, the USSR, in its recent manned space activities has already laid the groundwork for manned interplanetary flight. While it is impossible for Western observers to say for sure if Moscow has already decided to commit to an early (1990s) man-to-Mars effort, current and confidently predicted near-future Soviet space capabilities provide all of the building blocks needed to support that decision if and when it is made.

To whatever power opens up the Mars frontier, the planet offers unlimited opportunities for expansion. It provides a resource base for asteroid-belt exploration and exploitation, as well as hosting long-term human occupants who would gradually become settlers and colonists. Ultimately, the native-born human inhabitants of a future Mars may undertake large-scale climate modification, up to and including the idea of "terraforming" the planet (that is, rebuilding its atmosphere, oceans, and heat balance in order to open its entire surface to unprotected human occupation).

Key social and political issues are relevant to the decision to go to Mars, since such an effort would send ripples resonating through the entire home society. The role of exploration and of technological development (accelerated by such focused space efforts) in the physical, intellectual, and spiritual well-being of a society is discussed, with analogues to previous expensive, valuable, costly, priceless engineering efforts. A perspective is offered here on the plausibility of such an accelerated advance in space activities in an era (such as today) of slashing cutbacks in space budgets.

The implementation of a long-range program for the exploration and settlement of Mars can have profoundly beneficial spiritual and psychological effects on Earth's societies. It would truly be, and would be contemporaneously recognized as such, an epochal landmark in the entire history of the planet.

Scenario

"PEOPLE HAVE LANDED on Mars," the announcer intoned with a dramatic tremor in his voice that he had practiced for months at home in front of his mirror. "But—we won't know what shape they are in for another eighteen minutes, as their radio signals race back to Earth at the speed of light. Nevertheless, whatever is going to happen on this landing attempt has already happened and all that remains is for Earth to learn about it a few minutes from now."

Two dozen engineers near the announcer were peering intently at banks of consoles, lights, and display screens which reported on the health of a spaceship 200 million miles away and eighteen minutes in the past. The only noise was a quiet murmur of conferences relayed via earpieces and minimicrophones, economic terse code phrases in the lexicon of spaceflight jargon. "Houston, Ares," came a tinny voice through the static of the Air/Ground channel, "We're through high gate with 80 percent, and we're go." In the room, only one man was standing, like an orchestra conductor coordinating the brainpower of a hierarchy which stretched from Mission Control to other rooms across a continent and around—and off—the planet. He was the Flight Director, ultimately responsible for decisions being made now almost half a solar system away. At the most crucial moments in the long mission, when the alertness and advice of Mission Control were most acutely needed, they were helpless to affect the events that had already occurred.

Above the data screens in front of the room were at least a dozen digital clocks, some counting up, some down, some flashing, and some not changing at all.

"Here at Mission Control the mood is tense," the Public Affairs announcer continued. "The event clocks in front of the room tell the story. Three sets of times are given on the center console. First is the actual time on Mars (in Mars Mean Time), which has forty minutes more per day than Earth time. Next is the time that signals now arriving here in Houston originally left Mars so the flight controllers can follow their crew activities schedules. Third is the time on Mars that advice and computer commands sent from Houston immediately would arrive at Mars."

"The real-time clock now reads two minutes past the scheduled touchdown time. Signals sent this instant from Houston wouldn't reach the astronauts until twenty minutes after the landing. The astronauts will succeed or fail—indeed, have already succeeded or failed—at this crucial juncture without any more help from Earth." The speaker paused. "This is the Voice of Ares, Man-on-Mars."

Twenty minutes earlier, the four Ares astronauts were still straining against the two Gs of deceleration, twice Earth's normal gravity and the first such pressure in nine months. The commander (call sign, CDR), in one of the top two couches, prepared to stand up and monitor the final touchdown out the window. The life sciences officer (call sign, BIO) in the other top couch was ready to monitor the medical status of CDR. In the two bottom couches lay the geological sciences officer (GEO) and the flight engineer (FE) who watched complex displays, ready to call out either reassuring checkpoints or signs of deviations.

One of them, the FE, noted that the descent trajectory was right on the nominal track at the "keyhole" for main engine arm command. "Houston, Ares," FE called out, glancing at the propellant gauges, "We're through high gate with 80 percent, and we're GO." The GEO, meanwhile, was the only crewmember allowed to listen to the voice signals from Houston, so the other astronauts would not be distracted from crucial real-time conversations. If something urgent came up, GEO would relay the message or switch the voice from Earth onto the cabin loudspeaker.

Deceleration forces dropped gradually to less than half of Earth normal, as the falling capsule reached terminal velocity in the thin martian air: drag was just balancing Mars' weaker gravity, 38 percent that of Earth. The capsule's nearly horizontal entry flight path shifted in a wide arc toward the vertical, fifty tons of metal, fuel, and human flesh falling like a stone out of the martian sky. The air was too thin for a sonic boom or a contrail, so the visitors arrived unheralded and practically unnoticed—if anybody had been there to notice.

The CDR slowly sat up and swung his legs to the floor between the two sets of couches. *Here is where all that exercise had better pay off*, he thought, as stars swam in front of his eyes and he felt his heart pound. He inflated the G-suit pants which tightened around his legs and lower torso, keeping his blood supply from pooling too low to adequately support his brain.

His duty position in front of the window was only two quick steps away. Three more movements hooked his harness to overhead support straps and settled his elbows in two arm rests, positioned so that his hands firmly grasped two pistol grips which could assume control of the craft's autopilot.

"Ignition in fifteen seconds," called out the FE. GEO added, "Houston has just given us a go for the ignition and advised us to disregard any type-6 alarms which the failed 'B' chamber pressure transducer might trip off." BIO added a comment, "I'm upping your coolant flow a touch—your heart rate is good, for an old man!"

"Five seconds," FE called out, punching the ENABLE key on the console above. "Ignition enabled—three, two, one, mark." The ship shook and the CDR lurched in his harness as the G-forces rattled his teeth. FE continued, "I read four good main engines, no alarms. Gauges are tracking. Radar ranging data is GO. I have enabled 'accept' function. Coming up on low gate and looking good."

"Low gate status," called out the CDR, keying the transmit button to relay their words to Houston. FE was first, "Engines are GO, quantities are GO, navigation is GO." BIO was next, "Life support is GO, medical status is GO." GEO's turn came last: "Communications status is GO." The CDR summed it up for hundreds of millions of listening earthlings: "Houston, Ares, we are GO at low gate—so far, so good, guys."

The minutes passed without further extraneous comment. CDR would read critical navigation data reflected off the window, without moving his eyes from the landscape outside. The other crewmembers had also rehearsed their roles a dozen times on the long voyage up and out from Earth. They knew exactly what everyone else needed to be informed of as the landing neared, and communicated that information in a sparse, codified system that listeners would have thought both unintelligible and icy calm. It was neither.

The CDR's view of the surface was oblique, not straight down; directly below was blocked by the bulk of the spacecraft. But the visible surface gave him all the visual cues he needed to verify the autopilot's skill. A television screen over the window showed him the view straight down, and the CDR tapped his left control stick a few times to shift the predicted impact point out onto a smoother-looking area. *Now I can say I made a manual landing,* he mused, as the ground expanded in his view and on the screen.

And when the actual touchdown came, with a crash and a crunch that drove the CDR to his knees, the most general emotion of the crew was not the expected exhilaration or relief, but astonishment, so deep had been the feeling of "just another practice run." With no time for more than a brief cheer, they continued down their checklists, alert for any landing-induced failure which might call for an immediate takeoff.

The airlock was crowded with the CDR and BIO inside, each checking the other's suit and attached equipment. They barely had time to glance outside through the small porthole in the door. Just standing up was a struggle after so many months of weightlessness. *Thank goodness we don't have to carry the backpacks this time,* the CDR thought as he verified the attachment on his suit umbilical, which would allow him about fifty feet of radius outside the door. *I'll feel up to lugging it around in a few days, though,* he confidently promised himself.

"The mast television cameras show good deployment on both equipment bays," reported GEO over the suit headphones. "I've got positive talkbacks on all unlatch monitors." The stage was set. "And your ramp looks level with no dropoff or debris at the end."

Dropping the airlock pressure took about ten minutes, during which the two astronauts assured themselves that their suits remained airtight. On schedule, the CDR unlatched the hatch and pushed it open. The midafternoon Mars sunlight flooded into the lock chamber.

We've come a long, long way, the CDR thought as he stepped out onto the ramp and grasped the handrail. The brightness all around—or some inner emotion—brought tears to his eyes. *Well, I can't say I haven't had time to think about what I'm going to say,* he thought, recalling the months in the Mars-bound spaceship. His eyes drier, he took four quick steps and stepped down from the ramp, and opened his mouth to speak.

For the second time in a human lifetime, the entire planet's attention was taken away from worldly woes. Crime rates plummeted. Traffic congestion momentarily eased on the ground, sea, and in the air. Absenteeism soared; future birth rates were subjected to speculation. And a half-dozen minor but all-too-bloody wars paused for breath as participants contemplated an event even more momentous than killing: the arrival of life at an eons-dead world. Somehow, after such an event, the business of death seemed even more out of place on Earth.

Of the dozen Apollo moon-walkers, ten were still alive, to be besieged unmercifully by news reporters—many of whom did not even remember the actual landings a generation before. "Read the history books," they were told again and again.

For those on Earth old enough to have experienced and appreciated the first landings on the recently reoccupied moon, the déjà vu sensation was uncanny. They deliciously savored again the taste of triumph from their youth, along with the added pleasure of being able unexpectedly to share with their children the heady experience that had all too quickly been fossilized and dissected in the history books.

Worldwide exultation exceeded by far that of the brief Apollo era, tarnished as those years had been by unpopular wars and social stresses. Wars and stresses remained, but these first interplanetary footsteps were widely seen as a demonstration of new options for humanity. Far from being merely a rerun of Apollo, the Mars landings were perceived as permanently opening a new frontier, as breaking through old mental, physical, and spiritual barriers.

It was another giant leap for all humankind. Only this time, the planet's far-roving and high-reaching children were unlikely to come down to Earth again, ever, without remembering that the road to the stars was open.

Once the flight engineer's eyes got adapted to the dark, the view out the overhead viewing dome of the command module was dazzling. Tiny shooting stars flashed frequently against the bright, untwinkling lights of the stars. After half an hour of observation, FE had become convinced of the existence of at least two meteor radiants imbedded among the random paths of numerous "erratic" meteors, and entered a description into the log. *Two meteor showers active simultaneously*, thought FE. *What a falling rock zone!*

The photocells on the outer hull of the landed Mars spaceship tracked and recorded each flash of light for eventual computer pattern analysis. In a few weeks, the crew would emplace other

Art courtesy of Paul Hudson.

all-sky photo-cameras at sites up to fifty miles away, to allow triangulation and hence determination of the origins of the meteor swarms near Mars. Now, however, the first human being to watch shooting stars on Mars was fascinated by their beauty and profusion—and by a common bond. *Yesterday we looked just like that, a falling star above Mars, a visitor from another world. Just a temporary one, though!*

It was well past midnight on their first night shift on Mars. The other three crewmembers were asleep below, tossing restlessly against the harsh, rough feel of sheets after months of sleeping in midair. FE had napped fitfully in the afternoon and evening, and now had the duty of reviewing the volumes of incoming advice from Mission Control on the coming day's outside activities. There were the congratulatory messages from VIPs around the world, too. *On an alien world*, the astronaut thought, *somebody must always be awake—if only for the paperwork, and to answer the phone.*

In peripheral vision, FE suddenly noticed a new bright light far out on the horizon. With curiosity and a faint stirring of uneasiness, the watcher wondered at the brightness of the unmoving visitor. *It is even casting faint shadows,* came an awed realization.

The only human being conscious on the whole planet struggled with conflicting thoughts, anxieties, and desperate explanations. *Something alive is out there—we're not alone on this world.* Suddenly it all fell into place, and the FE laughed out loud. *It's Earth! Of course it is alive, and of course we're not alone here.* A glance at the clock verified that Earth was rising.

The faint murmur of the spacecraft equipment brought the FE out of the reverie: clicks, whirrs of fans, ticking of meters and pressure regulators, hisses from a headset lying on a nearby workbench. Before turning to the official duties, the FE one last time glanced at the meteor-streaked sky and the brilliant rising Earth. *We're a long way from home, and I'm glad to be here—but it will take awhile to feel at home on this world.*

We shall not cease from exploration
And the end of all our exploring
Will be to arrive where we started
And know the place for the first time.

T.S. Eliot

1

Rationale For a Manned Landing

IT IS PROBABLY safe to say that although enough is finally known about Mars for a safe manned mission to be planned, there are plenty of crucial unanswered questions remaining. The scientific questions, which would serve to outline a productive research program there, have yet to be fully defined; the local martian resources (water in particular) which could be profitably utilized by the visiting expedition have yet to be inventoried.

These tasks will require extensive earth-side analysis along with renewed unmanned Mars exploration in advance of the first manned expeditions. The big question is, How many more unmanned probes are needed, and what kinds of specialized research tasks are required, before the first manned visit to Mars?

The landing sequence for the Viking probes in 1976 was very suitable for robots, but not gentle enough for passengers. After a three-hour fall from orbit, the lander hit the outer reaches of the martian atmosphere and began to slow down. During a seven-minute period of deceleration, forces on the probe exceeded thirteen "Gs" before the module reached "terminal velocity," falling freely through the thin air. At 20,000 feet, falling at about 800 feet per second, the Viking popped an eight-foot diameter parachute which held for forty-five seconds and slowed the fall to about 200 feet per second. At an altitude of 4,000 feet, the parachute was cut loose and the craft's terminal descent engines were ignited. They operated for about forty seconds, throttled under control of a computer observing the approaching surface by radar altimeter. By the time the altitude reached zero, the descent rate had been reduced to a few feet per second—not too soon, and not too late.

Over the following months, both landers performed a wide array of specialized investigations. The Orbiters lasted for several years before running out of control gas; Viking-2, too far north for direct communications with Earth, was shut off just before its communications relay through the Orbiters broke, while Viking-1 continued sending signals (photos and weather data) for at least six years, its nuclear-powered instruments promising many years more of service.

The scientific results of the Viking probes have already filled several books, but many of those discoveries which are relevant to this book should be summarized. NASA chief scientist Gerald Soffen listed what he saw as the significant findings of the mission in 1978, and the following material is adapted from his summary.

Photography. Tens of thousands of pictures were taken, including some color pictures and some stereoscopic pairs. But less than 10 percent of the surface has been photographed at 100-meter resolution, and most areas were photographed only during one season, or even only once.

Infrared Thermal Mapping. Both permanent polar caps consist of water ice, to which dry ice (frozen carbon dioxide) is added and removed annually. At least several cubic miles of water ice are involved; the depth could range up to tens of feet in places.

Seismology. No major seismic events were detected by the sole operational seismometer, so Mars is much less seismically active than Earth. One quake measuring 2.8 on the Richter scale was detected at a range of seventy miles; unlike on the moon, which "rings" for long periods, quake signals on Mars are damped out very quickly, probably due to water and/or other volatiles trapped in the crust.

Meteorology. The air pressure varies seasonally by about 30 percent due to condensation of carbon dioxide at the polar caps. Summertime temperatures at the Viking sites ranged between a low of $-190°$ F. and a high of $-25°$ F. Carbon dioxide frost has been observed on the surface near the landers in wintertime.

Chemistry. Dirt at both sites was very similar; no rocks could be acquired at either site. The material was probably volcanic in origin, largely composed of iron-rich clay and oxides of silica and iron. Sulfur compounds make up about a tenth of the material, and magnesium compounds another tenth.

Molecular Analysis. Atmosphere was found to contain a surprising $2\frac{1}{2}$ percent of nitrogen. The ratios of gas isotopes suggest that Mars did not expel its gas from its early crust as efficiently as did Earth, but that the present atmosphere is only a small remnant (probably only 1 percent) of the original.

Biology. The highly complex instruments all gave readings which had been judged preflight to be positive signs of life forms. But a combination of other data, together with baffling behavior in some of the life-search runs, led most scientists to believe that a highly reactive surface chemistry was mimicking some types of biological reactions—although the chemical composition capable of such behavior has not yet been reproduced in earth-side laboratories.

There is a lesson for Mars exploration in some of those discoveries. It deals with the fact that there really is no such thing as "unmanned" space exploration. Rather, human scientists can send instruments to other planets, but they then must receive and interpret the data from their far-ranging emissaries. Exploration of Mars, in that interpretation, occurs primarily on Earth, by human beings.

And it is here that Mars exploration has suffered most in recent years. A lack of actual space probes is a highly visible symptom of a reduced national interest in interplanetary exploration, but far more damaging is the cessation of federal funding of earth-side scientific examination of the vast amount of Mars data already on hand, much of which has yet to be thoroughly analyzed.

Such a scientific process does not merely go into hibernation when the funding is cut off, to awaken again some years in the future when national priorities flip-flop once more. Such a resource—the people who have been experienced in receiving, processing, and analyzing the Mars data—dies when the teams of specialists are broken up and scatter to diverse, new careers. A gap between past and future is created, years lost and human resources wasted and, as usual, there is requirement to spend a far greater amount to someday restart the program than it would have cost to continue it at a reduced, modest, but reliable level.

It was to step into that gap that the Mars enthusiasts of the "Viking Fund" organized their fund-raising drive. Based in California, the grass-roots, pro-space private organization raised tens of thousands of dollars from the public to fund continued Viking data acquisition and analysis. In a precedent-setting legal coup, lawyers for the fund established that the money could not merely be dropped into the federal budget—legally, the government was compelled to spend it for the purposes specified by the contributors.

Perhaps it was only a drop in the bucket, but it was the right bucket. And the Viking Fund showed that a lot of ordinary people were still interested in Mars. Exploration continues, and so does a lot of dreaming about Mars, about exploration of Mars, and about a future human presence on Mars.

Following the success of the Viking missions, there were numerous proposals for follow-up investigations. Reports were published about rovers and rollers, orbiters and fliers, penetrators and scoopers, and drillers and sample returners. The variety of these proposals seemed to be testimony to the scope of the remaining human ignorance about Mars. NASA scientist James French put it this way in an article published in 1980: "The martian opportunities are so diverse that scientific opinion on the program options has not been unanimous. Despite these problems, a consensus may emerge." But it wasn't happening.

Perhaps that very variety of options, and the scientific community's apparent inability to reach an early consensus on the definition of one particular "next step," contributed to the demoralized retreat from new space goals that successively marked the Ford, Carter, and Reagan administrations. Whatever the cause, the result was that five years after the probes landed, they had become monuments to the *last* American Mars flights of the generation, perhaps of the century.

Things were no better in the Soviet Union. Bitter memories of the debacles of the early 1960s and the early 1970s, combined with tight space science budgets which saw the entire lunar-planetary effort concentrating on nearby Venus alone, led to a situation in which no new Mars probes were anticipated until the late 1980s at the earliest.

The scope of the available "next steps" to Mars may be reduced, and the options put into priority, by use of the mental discipline connected with a single main goal: support the planning for a manned Mars expedition. Those questions crucial to manned operations must be answered first; those questions which can be answered better by manned systems can be put off.

The Apollo analogy may be helpful. Unmanned probes in the Surveyor lander and Lunar Orbiter programs preceded the manned landings by two to three years. Although they did carry some scientific instruments, their main purpose was to scout the lunar surface and environment, and map prospective landing sites, thus clearing the way for manned activities which actually carried out scientific programs far more sophisticated than would have been possible for the unmanned probes.

A definition of four classes of unmanned Mars exploration modes was given by James Cutts of the Planetary Science Foundation in Los Angeles, a NASA contract "think tank." Cutts, who came to Boulder in 1981 to discuss landing site selection, outlined the following modes: Orbital Science, from space; Network Science, involving several surface sites; Mobile Surface Science, involving a robot rover; Sample Return Science, using earth-side laboratories to study Mars material brought back by automated equipment.

With a stronger emphasis on manned operations, the Boulder conference modified that list significantly. In place of a generalized orbital science program, analysts specifically recommended a focus on "water mapping experiments." The network concept was dropped. Sample return became a support for planned Mars mining, the processing of local materials for air, water, and fuel (and hence need not precede the first brief manned visits). And one entirely new class of operation was added: "High Resolution Imaging," possibly involving an orbiting photo-robot which dispatches film cannisters back to Earth for detailed processing.

The resulting menu and price tag constitutes a fascinating strategic problem involving assignment of limited resources against uncertain targets with poorly estimated payoffs—a job for intuition, not mere cost accounting. Here are the options in detail.

Orbital Science. According to Cutts, orbital science consists of "many different observations from orbiting spacecraft (e.g., imaging, spectral mapping, gamma-ray mapping, aeronomy, gravity, magnetic, etc.), attacking a range of global, whole-body, surface, and atmospheric science questions." A spacecraft to accomplish all of this would probably weigh about 10,000 pounds en route (requiring one Space Shuttle launch) and 2,000 pounds in low Mars orbit; it could cost between 300 and 400 million dollars. The reflectance spectrometer on board would gather data on rock types, soil composition and aging, and other surface-related data characterized by the color variations in reflected sunlight. The gamma-ray spectrometer could determine some elements in the rocks from subtle radioactivity. A camera system could cover some regions down to a five-meter (sixteen feet) resolution, much better than Viking (good enough to spot houses on the surface, but not to spot any cars in the driveways!). A radar altimeter would measure the precise shape of the planet, using off-round variations in that shape to estimate crustal thickness and plasticity, and hence thermal state and evolution. Other instruments would survey surface temperatures and air pressure.

All of these items are of great scientific interest, but the Boulder conference split them into two groups: nice and vital. The only vital parts were those concerning the surface distribution of water, because of its use in life support, and high-resolution photos to ascertain the roughness of candidate landing sites and to prepare advance traverse maps of the areas for exploration. (In the latter application, even the proposed orbital imaging system would not be good enough). Besides, such orbital science surveys might have some inherent limitations. A 1977 study by planetary geologists at the NASA Johnson Space Center (which, for convenience, I'll refer to as the "Bogard Report," although it represented the work of at least a dozen specialists) issued a cautionary warning on the design of an unmanned orbiting survey probe: "Multispectral imaging and reflectance spectroscopy could provide important data on the distribution of surface materials," the report admitted. "However, these spectra may be dominated by dust . . ., which is suspected now on the basis of Viking to be similar in composition all over the planet." This is unlike the moon where windblown global mixing never occurred.

In the place of the generalized orbital spacecraft, a specific "Water-Sniffer" probe was proposed. At Boulder, Dr. Charles A. Barth, director of the Laboratory for Atmospheric and Space Physics at the University of Colorado, described his proposed small (1000-pound), specialized, and inexpensive Pioneer-class Mars orbiter. "Specific objectives," he wrote, "are to measure water on the surface as frost, in the atmosphere as vapor, and in the clouds as condensates." The instruments, which would observe both re-

The "water-sniffer" probe, to chart the water resources of Mars, is an urgently needed and generally agreed-upon next step.

flected light from below and blocked light as the sun rose and set across the horizon twice each orbit, would operate at various specific wavelengths in the infrared and ultraviolet, where certain absorption bands provide the sought-after data.

Network Science. According to Cutts, this would consist of "systematic, long-duration observations (e.g., seismic, meteorologic, chemical, imaging, heat flow, water detection, magnetic, etc.) at several points widely distributed over the planet." Since the cost of a fleet of Viking-class soft landing vehicles would be prohibitive, the network sites could instead be established using tough "hard landers" or even tougher, remarkable vehicles called "penetrators."

A penetrator is simply a torpedo-shaped structure that drops from space, uses a small parachute for descent stabilization, and then impacts the ground at high speed. One section is detached at the surface, while the arrow-shaped forebody plunges into the solid ground (dirt or ice or even rock) to depths of tens of feet; the forebody experiences a G-force between 200 and 2000, while the afterbody, which remains attached via a reeled-out wire, experiences up to 20,000 Gs.

Those incredible stresses are survivable for certain types of solid-state electronics and well-anchored mechanical equipment. In fact, penetrators have been extensively tested for use on the moon, Mars, and other planets. Penetrators have been widely used in military operations such as secret ground-monitoring equipment, dropped behind enemy lines to listen in on noises of motorized traffic. Since they are so firmly attached to the ground, they make first-rate seismic detectors. They also can find—by actually reaching—buried permafrost zones which are generally believed to exist in many areas of Mars.

One orbiter craft can deploy up to a dozen such probes, each weighing only about 200 pounds while still attached to the spacecraft. After deployment, the orbiter must act as radio relay and can also act as an observation platform for other investigations. Thus a penetrator network deployment could be combined with an orbital science mission. Cost estimates range between 150 and 250 million dollars for such a network mission.

Although the network proposal was not judged essential by Boulder conferees as a preliminary to manned missions, its role as

Soviet mobile lunar vehicles, the unmanned Lunokhods, operated on the moon in the early 1970s.

an on-site detector of buried permafrost did earn it an evaluation of "highly desirable," especially if other sources for water are not more easily detected.

Mobile Surface Science. According to Cutts, this mission could provide "quite detailed and complex investigations (e.g., chemistry and mineralogy of sequences of rocks and surface material, atmospheric chemistry and isotopic composition, biological experiments, geophysical experiments, surface properties, and soil mechanics) at a number of surface locations within a limited mobility range." A wide variety of hardware could satisfy this role: small, tethered sample-gathering systems could range out 100 feet from a landed laboratory on umbilical cords; a large, autonomous rover weighing 1,000 pounds could range more than fifty miles, travelling a quarter mile per day, under semi-internal computer control; a giant "beach ball" with motorized center-of-gravity offset drive (or wind power) could roll across the surface, stopping to extend instrument booms out through special portholes; a 500-pound aircraft powered by hydrazine engines could fly at about 200 miles per hour for up to thirty hours, making photographic, magnetic, gravitational, and geochemical measurements, and with special engines could even land and take off again, gathering samples.

The Boulder conference allowed that such rovers were desirable, but possibly not worth the cost. Cutts' own report gives this explanation: "It appeared to many planners that a mobile Mars laboratory with a 100-kilometer range could both extend scientific knowledge of Mars and capture the interest and enthusiasm of the public. When the unexpectedly high costs of a mobile laboratory mission became known, the concept of mobile science on Mars rapidly fell out of favor . . ." Although the report voiced the common frustration of being able to see the horizon from the Viking sites but not being able to satisfy the human urge to look over the next hill, he cautioned that "designing a system of planetary exploration for the primary purpose of relieving these frustrations does not necessarily result in major scientific advances. It may turn out that the vista viewed from one martian ridge is much like every other, and that once all four sides of a martian boulder have been scrutinized in detail, the public's appetite for such activities is sated, not whetted."

The experience of the Soviet "lunokhod" rovers on the moon would be more valuable if it were not so distorted by Moscow spokesmen eager to put the best light on every Soviet space shot. In fact, only two such rovers were ever successfully landed, in 1970 and again in 1972, and despite the claim that they represented a breakthrough in solar system exploration technology, the program was soon terminated. Privately, Soviet lunar scientists conceded that the rovers were far too expensive for the minimal scientific results, and that controlling them was far more complex than originally anticipated. One, in fact, probably fell over in a gully, although Moscow never admitted as much. Such experience might explain the current lack of Soviet enthusiasm for the once highly publicized "marsokhod" Mars rover vehicle.

Sample Return Science. This would involve, in the words of the Cutts Report, "the return of rationally chosen samples from carefully chosen areas (with mobility and analytical capability for sample selection and acquisition) for detailed chemical, biological, physical, geological and chronological studies." It would be the most expensive and complex unmanned Mars expedition ever, costing upwards of $2 billion. In Jim Cutts' estimate, this figure may be as little as two or three times lower than the cost of a manned mission, or as much as ten or fifteen times cheaper, in the estimate of Louis Friedman, president of the Planetary Society. It would arguably be the most scientifically rewarding as well.

Up to 20,000 pounds of hardware would depart for Mars requiring two Shuttle launches to carry the spacecraft and its booster stages; the return would be a cannister of Mars soil, weighing (cannister plus soil) about three or four pounds.

At the minimum, the system would deliver a single, grab-bag-type, surface sample. This is not desirable compared to much more elaborate systems that are only slightly more expensive. In the most highly desired mode, the hardware could provide subsurface sampling, site documentation, sample selection and pre-launch preparation. It could also have some reasonable mobility to pick

The Mars sample return mission is the most sophisticated unmanned spacecraft ever proposed.

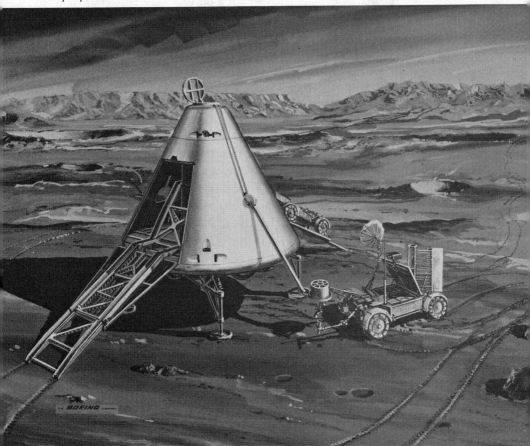

and choose rocks near the lander. Perhaps even a separately launched mobile rover would spend several months traversing the surface, filling up a cookie jar with material for transport back home, but that would add complexity and the special problem of *surface* rendezvous.

The need for sophisticated sample selection has been underscored by a comparison of Soviet and American moonrock sampling technologies—automated versus human. In 1977, Dr. Bevan French, a leading NASA lunar geologist, wrote:

> There is a continuing debate over whether the scientific results from the Apollo program could have been obtained at much less cost with unmanned samplers . . . ([which] returned only a small amount of soil and no large rocks). . . . In many cases the answer is no. . . . The larger samples returned by the Apollo missions were essential to learning about formation and exposure ages, highland breccias, microcraters, lunar magnetism, the nature of solar wind, cosmic ray particles, and the layering and history of the lunar soil. . . . Without the large rocks from Apollo, on which the true formation ages could be determined, our whole view of lunar history might have gotten off to a very false start, and we might never have learned that the moon had been an active, evolving planet for a billion and a half years.

Of course, future unmanned systems can take that caveat into consideration. Additionally, analysis technology has already advanced significantly (partly, in fact, under impetus from the need to analyze extraterrestrial samples). According to the 1977 Bogard Report, "Individual igneous rocks as small as four millimeters [one-sixth of an inch] diameter have been studied successfully by a wide variety of techniques including petrology, chemistry, and age analysis. . . . The definition of a rock has been extended by the lunar program" to include such small flakes. In fact, noted the NASA-Houston study, "Rocks larger than three to four centimeters [an inch or two] begin to exceed the optimum size for geochemical-petrological studies . . ." so a sampler, manned or unmanned, should be able to chip off fragments of the desired optimum size.

Scientists at the Boulder conference were also eager to see Mars samples in advance of manned expeditions, but primarily this would be in order to analyze the material for utility in supporting

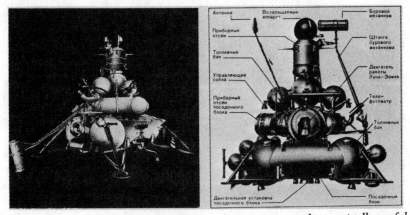

Soviet unmanned moon samplers of the early 1970s were only marginally useful and were much less scientifically productive than manned Apollo flights.

the manned expedition as a source of air, water, propellants, or other usable materials. As Jim Cutts summarized at the conference, "*If* we will depend on local material, it's crazy to go without seeing the dirt."

The sample return controversy was put into perspective at Boulder by a discussion from Dr. Ben Clark, who helped build the Viking probes. "The main purpose is to bring back science, not rocks," he pointed out, alluding to the crucial sample selection process. "People can pick up thirty or forty pounds of material, selected to be representative, that could teach more than a robot return cargo of a thousand pounds." Sampling experience on Apollo underscored Clark's point, since the most valuable samples were those selected by human judgment but which a robot (even with advice from Earth) might have overlooked.

One other type of object might make the long, return journey from Mars: photographic film, which provides far higher resolution of surface details than data sent back as radio telemetry. An orbital mission lasting several months could scan interesting areas, load the film in a return cannister, and fire it off on its way.

The value of such high-resolution imagery was underscored by the experience with selecting the Viking landing sites. Although orbital photo-analysis and radar data from Earth both promised that the Viking sites would be smooth and safe, in fact they turned out to be neither: post-flight analysis based on actual distribution of big boulders around the landing craft showed that the odds against a safe landing in either area were at least two to one! In other words, both Vikings had "lucked out." To avoid such unpleasant surprises in the future, much better photographs of selected landing zones across the face of the planet must be obtained prior to the manned landing.

As with the sample return, the mission sequence and the need for unmanned missions at all depends on overall strategy. Assuming the first manned landing does not count on using local resources for exploitation, which is likely, then logic no longer requires an *unmanned* sample return: the first manned landing could fulfill that role for the later landings planning resource extraction. Similarly, high resolution images are needed before the first manned landing, and if that is the first manned flight to Mars, then the photo mission must be an unmanned mission. But there are scenarios, as we shall see, in which the first manned mission to Mars could merely be a Phobos-Deimos visit, during which the requisite landing-site photography could be accomplished by the astronauts as an additional function.

These possibilities led the Cutts Report to raise a serious question about their ultimate utility: "The problems of future Mars exploration with unmanned vehicles," it warned, "are exacerbated by the fact that manned exploration of the planet will be feasible technologically (though not necessarily economically) within a decade of an unmanned sample analysis mission. The potential of a manned mission for scientific investigation is so much greater than an unmanned mission that it is legitimate to question the need for an interim, high-cost sample analysis mission. On this view, the development of sophisticated unmanned sample analysis missions for Mars, whether they be surface laboratory missions or sample return, constitutes a technological and scientific blind alley and a capability that is superseded so soon that it can never be adequately exploited."

What is the bottom line about future unmanned hardware choices? The water-sniffer probe (perhaps a good program name is "Prospector," or even "Bedouin," but it will probably just be called "Pioneer-Mars") is evidently crucial, and should be launched as soon as possible to aid in follow-up planning. Other probes, such as the rovers and penetrators, are not critical to the preparation of a manned expedition. The fate of the automated sample return mission proposals, with their great cost and complexity, depends on the ultimate speed of the manned expeditionary planning; where the latter is admittedly much more expensive, it may turn out to be a lot more feasible in terms of politics *and* produce much more real science as well.

A manned mission, after all, can carry out practically all of the highly varied missions proposed for all of these extremely diverse unmanned options. That thought was best expressed by the far-sighted space systems planner Krafft Ehricke, who in 1970 wrote the following. "By increasing the scope of the manned mission . . . the return grows faster than the mission cost. This eventually makes the manned-mission mode superior in cost-effectiveness (e.g., dollar per useful bit) to the unmanned mission. It also enables the manned mission to perform investigations which the unmanned vehicle could not do."

Options for unmanned missions to Mars should be considered in that perspective.

*It is difficult to say what is impossible. The dream of yester-
day is the hope of today, and the reality of tomorrow.*

Robert H. Goddard, 1901

2

Getting There: Spaceships and Propulsion

SPACESHIPS

If all of the different types of Mars spaceships proposed in the last thirty years were actually built and then laid end to end, the resulting contraption might reach halfway to the intended destination. Designing Mars vehicles has been a sort of engineering thumb twiddling for many, many space-hungry engineers, and for a few happy years there even was a chance, a genuine chance, that some such spaceship would really be built.

If ever such a chance is to come again, there is good reason to survey the major plans, and that is to extract any philosophies, ideas, and conclusions from these designs which might still be valid. Most of the technology has become either obsolete or obviously impossible, but there still remains, with careful sifting, a residue of valuable work and really useful suggestions.

The first engineering analysis of the actual hardware needed on a man-to-Mars expedition was published in 1952 (half a century after science fiction had irrevocably opened the Earth-Mars highways) by Wernher von Braun, the expatriate German rocket engi-

42

neer who, with his team of former Peenemunde V-2 specialists, was now working at the White Sands Missile Range in the American Southwest. Entitled *Das Marsprojekt* in its German edition, it was also published in translation by the University of Illinois Press as *The Mars Project*. (It was a very limited edition: even the NASA library in Houston has only a photocopy of the book.)

Ten spacecraft, each weighing 8 million pounds and carrying seven men, were to be assembled in low earth orbit (a few hundred miles up, just above the atmosphere), for a convoy mission to Mars. Three of the ships carried 200-ton "landing boats" capable of landing fifty people and 150 tons of supplies and equipment on the martian surface for a 400-day visit.

To assemble this interplanetary armada, von Braun foresaw a fleet of fifty reusable ground-to-orbit shuttlecraft weighing 25 million pounds and carrying 80,000 pounds of payload, each rocket flying every ten days for the eight months it would take to assemble the fleet. A modest price tag of $50 million (1953) per ship was based on fuel costs alone; von Braun offered the highly optimistic assertion that "the logistic requirements for a large, elaborate expedition to Mars are no greater than those for a minor military operation extending over a limited theatre of war"—a flippant comment from a man whose V-2 project strained the economy of wartime Germany (and in the final analysis probably saved tens of thousands of lives by absorbing resources that otherwise would have gone into far more effective killing machines).

On later reflection, von Braun pared down the scope of his proposal. In the 1956 popular book, *The Exploration of Mars*, (as co-author, science popularizer Willy Ley), the fleet was reduced to only two ships carrying twelve men each. The ships weighed 360,000 pounds each. Meanwhile, the ground-to-space launch vehicle intended to transport the ship's pieces and fuel into orbit was also scaled down to carry only 22,000 pounds per launch, but this still required 400 supply launches over a seven-month period, or two launches per day. The idea of recovery and reuse of the launchers was discarded. "It appears quite possible that the costs of salvage might exceed the savings," the authors wrote. They were to be proved quite correct for twenty years until technological progress finally made the Space Shuttle marginally economical.

The 1950s also saw other extremely ambitious Mars expedition plans which were a strange combination of wrong assumptions, overlooked essentials, and remarkably prescient suggestions. Felix Godwin's 1959 scenario involved five ships, each weighing 1.2 million pounds and carrying twelve crewmen (Godwin specified the crew specializations and duties in excruciating detail). Although his plan was based on a too-high air pressure estimate (eighty-five mbars versus the actual value of six to eight mbars), other suggestions were right on the money: the use of a lunar electromagnetic cannon (now called a "mass driver") to send moon ore into space; propellant manufacture on Mars (using the wrongly expected "abundant nitrogen"); air-drag braking during the Mars arrival maneuver; surface gardens for food supplies; and ultimately the "large-scale improvement of the planet" (i.e., terraforming). The most remarkable feature about Godwin's scenario (his estimate of space vehicle weight and size was as close as anyone else ever got to modern concepts) was that he wrote and published it when he was nineteen years old.

It was not long before the concept of Mars-bound ships using liquid-fueled or nuclear-powered rockets was matched by another propulsion concept: ion drive, or electric propulsion. This involved very long-duration, but very low-thrust engines instead of the brief-action but high-thrust conventional rockets. (Technically, there are at least two types of electric propulsion: ion drive, and a recently gaining alternative called magneto-plasma dynamic or "M.P.D. plasma.")

Ernst Stuhlinger, one of the V-2 engineers, championed the idea in the late 1950s and afterwards. A scenario in 1961 (the year the first astronauts and cosmonauts ventured briefly into space), involved five three-man ships weighing 800,000 pounds each, for a 572-day round trip including a month on Mars (but later, the required weight of the nuclear electrical power plant had to be significantly increased). By 1966, Stuhlinger was offering a refined plan for a million-pound ion-drive space vehicle, which involved a four-week visit to Mars by a 125,000-pound landing craft, while the low-thrust mother ship slowly spiraled into a low enough orbit to make a successful pickup of the returning astronauts. With reasonable assumptions about power plant weight, ion-drive advocates claimed that their type of vehicle could perform equivalent-duration Mars missions at *half* the launch weight of nuclear-thrust space-

TVC-755

In 1957, Walt Disney produced a remarkably prophetic film, "Mars and Beyond," showing a fleet of ion-drive vehicles carrying winged landing craft for Mars exploration and ultimate colonization.

craft—a claim which remains valid to this day, stymied only by the failure of anyone to finance the development and testing of large ion-thrust engines for space missions.

In the mid-1960s, serious engineering studies were undertaken under the premise that the Apollo man-on-the-moon program might lead to a similarly funded man-on-Mars program in the following decade. In a baseline mission using chemical propulsion for Earth departure, Mars arrival, and Mars departure, the necessary vehicle weighed more than 5 million pounds in its parking orbit just above the Earth's atmosphere. Using air-drag braking at Mars, the vehicle would weigh only one-third of that amount; adding nuclear propulsion for the Earth departure would cause another reduction of 65 percent of the new figure, or only one-fifth of the original all-chemical figure. Analysis also showed that a man-to-Mars expedition which only intended to orbit (or merely visit the two small moons of Mars, Deimos and Phobos) would weigh up to 75 percent as much as the lander mission, regardless of propulsion system. These relative values represented genuine engineering relationships and are still valid.

Probably the best blueprint for a Mars-landing manned spacecraft, the MEM, was developed by the North American Rockwell Corporation in 1967, in a NASA-sponsored study headed by Geoffrey S. Canetti. The MEM was seen as a $4 to $5 billion (1967 value) project, including flight vehicles and an extensive unmanned and manned test program.

The Canetti team's MEM weighed 80,000 pounds and carried four crewmen from Mars orbit to the surface and back. The nominal surface staytime was thirty days. An Apollo-type atmospheric entry shape was preferred for aerodynamic reasons. After landing, the crew would live in a cabin located near the surface, between propellant tanks used for the descent engine system. (It was not proposed, but it is possible, that they could also inhabit some of those scrubbed-out tanks, too!) The ascent system was a single-stage rocket with a small cabin on top; it was powered by a combination of FLOX (fluorine and liquid oxygen) and methane, two propellants chosen for their safe storability over nearly a year in space before engine firing.

One of the most intriguing aspects of this report was the detailed testing plan. Since this MEM design remains the most reasonable one ever developed, a very similar test plan will almost certainly be followed whenever the new version of this 1967 MEM is finally built by the children and grandchildren of the engineers who first designed it.

First would come an unmanned downrange launch to test the heat shield in Earth's atmosphere. Then an unmanned vehicle would be sent into orbit (a three-unit SRB-X design should do the trick, as described in a coming chapter) for a 200-day coast followed by an unmanned earth landing (the heat shield good enough for an Earth landing would weigh 1,300 pounds more than one just barely good enough for Mars, but it would reduce the testing program cost by a full billion dollars). Additionally and in parallel, a fully configured unmanned ascent stage would also be launched aboard a booster, and would be fired at high altitude, perhaps going into orbit.

Manned tests would come next—if you can't trust the hardware near Earth, you have no business trusting it 100 million miles away. First would be a MEM entry and landing from an initial parking orbit (carried there aboard a large-diameter SRB-X booster since its 30-foot-diameter heat shield is twice as wide as the Shuttle cargo bay); the crew would rendezvous with it in orbit, as passengers aboard a Space Shuttle, and transfer into it for the landing back in some desert in New Mexico, Tibet, or the Andean Altiplano. Next would come a manned ascent to orbit aboard a launch booster (SRB-X again?) plus a MEM ascent stage as upper stage, to be followed by a mock rendezvous and docking to a waiting Space Shuttle (which would retrieve and return the astronauts). Additionally, according to the Canetti Report, modified balancing and cross-piping of the MEM's propulsion stages plus auxiliary braking rockets would allow a manned MEM to make a practice "dry run" landing and liftoff on the moon.

One important observation was made about the possible need for near-Mars testing of the MEM prior to the actual manned landing mission. The Canetti Report was fairly definitive. "Unmanned MEM Mars flight tests using unmanned spacecraft, manned

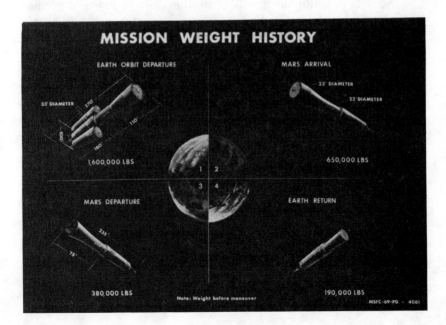

MISSION WEIGHT HISTORY

EARTH ORBIT DEPARTURE   MARS ARRIVAL

33' DIAMETER   270   110   33' DIAMETER   22' DIAMETER

1,000   140'

1,600,000 LBS   650,000 LBS

1   2

3   4

MARS DEPARTURE   EARTH RETURN

235'   75'

380,000 LBS   Note: Weight before maneuver   190,000 LBS   MSFC-69-PD - 4061

orbiter, or fly-by missions are not necessary. . . . Additional con-
fidence to be gained by an unmanned MEM Mars flight test program does not at this time appear to justify the increased cost and
schedule constraints which would be imposed on the manned Mars
landing program." At least, that's how it looked even before Apollo
had landed on the moon—and that's probably how it still looks
today.

All this talk about "millions of pounds" in parking orbit probably does not mean much if you don't have any space operational
perspective, so here it is. In 1980, for example, the USSR launched
1.2 million pounds of usable hardware (and almost another million
pounds of spent rocket stages) into low Earth parking orbit atop
90 boosters. In 1985, NASA plans to fly twenty-four Space Shuttle
missions, which would involve putting 6 million pounds of material
into orbit, with an equivalent *payload* weight of more than 1.5
million pounds (much of it in the form of propulsion stages to push
the actual satellites into higher orbits). And in 1981, the Pentagon
confirmed long-standing rumors that the Soviet Union was developing a giant space booster with a payload capacity in excess of
400,000 pounds; three launches of such a rocket could easily assemble a million-pound man-to-Mars spacecraft.

So it is not utopian to talk of millions and millions of pounds
of payload for a Mars expedition—in fact, since it merely anticipates

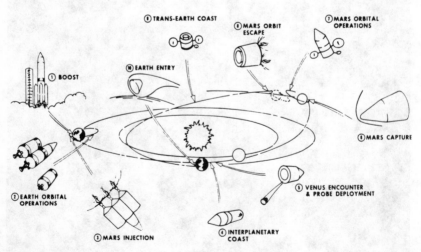

Mission profile to Mars and back.

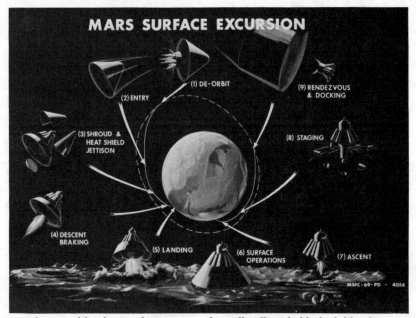

The actual landing and return to orbit will still probably look like this.

that national launch capability will only double in ten or twenty years, it's probably a pessimistic and overcautious estimate. The cargo costs *will* be high—but the common carriers will be there, willing to sell the ticket, and able to deliver the goods.

By the time of the 1969 Apollo moon landing, space engineers had developed a "NASA standard man-to-Mars" mission. For many space workers, the moon was only a detour, a temporary distraction from the serious business of getting to Mars—and numerous groups within the space industry had spent years getting ready for the next "big push." The way they generally conceived of it was something like the following.

One Saturn-V launch would carry the Mars spacecraft itself, about 226,000 pounds. Up to five more Saturn-V launches each would carry a nuclear propulsion module. Several smaller Saturn-Ib launches would carry astronaut crews both for orbital assembly of the pieces, and for the actual mission.

Three propulsion modules would be dropped after leaving Earth (that's half the initial weight of the vehicle). One more would be expended to achieve orbit around Mars after eight months of interplanetary coasting.

The Mars Excursion Module would descend to the surface, leaving some of the crew in the still-orbiting Mission Module. Atmospheric drag would slow the MEM most of the way, aided by parachutes and braking rockets in the last few minutes of the descent.

The expedition would remain on the surface for weeks, perhaps months, depending on the planned return route to Earth. Astronauts would range over hundreds of miles and collect half a ton of carefully selected and documented samples and drill cores.

At the end of the surface stay, the crew would load their results into the ascent stage and blast off. Halfway up, emptied propellant tanks would be jettisoned; once in orbit, the spacecraft would seek out and dock with the waiting Mission Module. The crew and their samples would be transferred to the main ship and the ascent stage

The landing would commence from the Mission Module after the main spacecraft went into orbit around Mars.

MARS
INITIAL LANDING

MSFC-69-PD

A small rocket would be used to blast the ascent stage from the surface to the orbiting Mission Module.

would be cast off. With the remaining propulsion stage, the Mission Module would head back for Earth, many months away.

On the last day of the mission (two or three years after leaving Earth in the first place), the crewmen would enter a small landing capsule and separate from the expended Mission Module. They would then hit Earth's atmosphere (perhaps with some braking from a remaining small rocket module) and either land directly, or steer into a parking orbit to await pickup. The man-to-Mars-and-back expedition would be over.

Variations to this mission sequence included the suggestion that two identical spacecraft be sent out together to provide mutual assistance—a concept that dates back at least as far as von Braun's 1952 proposal. There he had written, "Columbus . . . chose not to sail with a flotilla of less than three ships, and history tends to prove that he might never have returned to Spanish shores with his report of discoveries had he entrusted his fate to a single bottom. So it is with interplanetary exploration. . . ."

In disagreement, a study at the NASA astronaut center in Houston (where contingency planning and crew safety had been honed to a fine art) did not believe that the cost of a second launch was justified by any increase in mission safety. The report, supervised by mission-planning engineer Morris Jenkins, concluded that convoy operations are *not* the best way to assure crew safety. Wrote Jenkins in 1971,

> The advantages of the mission convoy technique are questionable. Simultaneous propulsion maneuvers are required if convoy rescue is to be consistent with tolerable propellant requirements. Furthermore, should one of the spacecraft fail to obtain orbital inser-

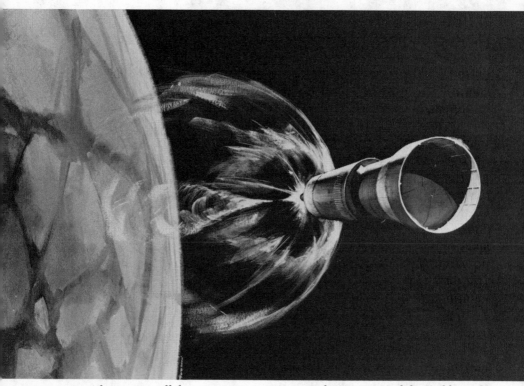

After casting off the now empty ascent stage, the Mission Module would use the remaining propulsion stage to head back to Earth.

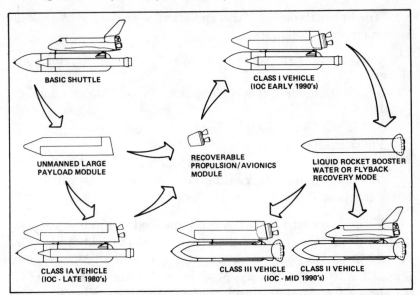

The Space Shuttle will form the backbone of ground-to-orbit transportation for the next quarter-century. Evolutionary designs include heavy cargo carriers as powerful as the Saturn-V series. Four or five such launchings could be adequate to assemble the Mars spaceship in parking orbit 200 miles up.

tion, a rescue effort would probably not be possible; or, at best, would require extremely complex operational techniques. The fact that rescue is not always possible, together with the fact that reliability standards could not be relaxed on either spacecraft, led to the exclusion of the mission convoy concept. . . .

The alternative is to make the single spacecraft into a "double vehicle," with dual independent life-support systems each capable of supporting the entire crew; dual propulsion systems (which normally operate in tandem but could operate alone if needed) also may be reasonable.

Another valuable analysis contained in the Jenkins Report dealt with crew composition. If only a few astronauts (five in this case) could make the trip, what kinds of skills should they represent and how should they be cross-trained to provide crucial skills in the event of the loss of one or more members of the crew?

The five seats on the Mars expedition, as envisaged by Jenkins, were defined as follows:

Commander (backup as medical technician and systems engineer);

Systems Engineer (backup commander and Mission Module pilot);

MEM pilot and geologist (backup systems engineer);

Physician (backup geochemist);

Geochemist (backup medical technician and MEM pilot).

(Actually, Jenkins tagged the mission scientist as a biologist, but a geochemist seems more logical today.)

Various minority reports were published in those years, suggesting clever ways to avoid the massive initial weight of the Mars vehicle in parking orbit. One proposal removed all Mars-vicinity propulsion from the Mission Module, and instead sent a beefed-up Mars Entry Module ahead as Mars neared, for a brief landing and reconnaissance before a desperate launch to catch up with the passing Mission Module, which would not enter a Mars orbit at all. Another proposal saved launch weight by *not* including a return ticket: the crew would land on Mars, set up a base camp, and await pickup by a follow-on expedition two years later.

By the mid-1970s, man-to-Mars was a taboo topic among government, university, and industry space officials. So, as had happened a quarter-century earlier, private groups and even dedicated individuals produced their own, often highly original, mission plans. Making use of Apollo and Viking experience plus the general advance of aerospace technology, they charted new routes to Mars, and designed new spaceships. And among all these plans are bound to be ideas useful when the time comes again to officially plan a NASA man-to-Mars expedition. If not, at least there is little doubt that their plans have been carefully translated into Russian.

A private California group called the "World Space Foundation" (many of whose members are scientists and engineers in the space industry) developed its own man-to-Mars plan in 1979 and 1981, based also on solar sails and the concept of sending supplies ahead on a slower but more efficient transportation system than that used for the follow-on crew transfer. The forerunner ship weighed one million pounds in parking orbit and took four years to get to Mars using solar sails; the manned ship weighed only 520,000 pounds before departure. This very low weight was possible by use of a risky maneuver in which the astronauts would have to make a rendezvous at Mars with the supply cache in order to get home alive, since their original Earth-Mars ship was a one-way stripped-down vehicle with no return capability.

Other reports kept popping up. Robert Parkinson of the British Interplanetary Society was championing a plan in 1980 which called for using Spacelab-type modules for a joint American/European expedition. Leonard David, editor of the National Space Institute's newsletter *Insight,* wrote a series of articles expounding on the growing feasibility of manned Mars exploration, as did noted space author Harry Stine. The volume of ideas was reaching critical mass—and all in the private sector, without encouragement from either government or industry space officials.

The fruit of this intellectual ferment and output was a conference in Boulder, Colorado, in the spring of 1981. Sponsored by an energetic and imaginative group of graduate students at the Laboratory for Atmospheric and Space Physics, the conference was supposed to unite all of the different groups who were independently working on similar problems, so they could share their insights and encourage each other in the face of their difficulties.

As news of the conference circulated in the space community, dozens of people stepped forward, often almost literally "out of the closet," with major man-to-Mars ideas and studies. They spent three days at the base of the Rocky Mountains describing their own work and running workshops on particular problems associated with the goal of sending men and women to Mars. And they came away with more than new facts and analyses. They came away with the realization that a "Mars Underground" is thriving at universities, industrial centers, and government space laboratories, both military and civilian, an underground formerly disconnected and mutually ignorant of each others' works. No longer. The Boulder conference of 1981 established links for a new (and necessarily unofficial) network of Mars spaceship designers eager to cooperate on a synergistic effort to prove the feasibility and desirability of a manned expedition to Mars at the turn of this century.

PROPULSION

Of all the far-out revolutionary options, only light sails and electric propulsion seem possible candidates for turn-of-the-century manned Mars expeditions. I hope I have to eat my words, and there are a few specialists who have promised to make me do so, but any breakthrough would require such major funding that all reasonable cost estimates for the expedition must be thrown out the window.

At first glance, the solar sail concept looks like magic: a craft so-equipped allegedly could fly across the solar system at essentially no cost in fuel. The photons from the sun's radiation, and not the ions from the "solar wind," provide the propulsive force to billow the giant metal-foil sail. Since the thrust is proportionate to the area, while the acceleration is inversely proportionate to the total system mass, the sail-powered spacecraft must be as lightweight as possible for as wide a sail as possible. This has led spaceflight theorist Eric Drexler to suggest the poetic name *light sail* for the technology in question—an extremely appropriate suggestion.

But light sail technology, while highly promising, remains far from proven in practice. In the late 1970s, during a NASA study competition between sail and the more familiar (and already flight-tested) electric thrust for a Halley's comet rendezvous, the sail lost out for this reason (and the electric system later lost out for budgetary reasons). Hence no official studies are going on anywhere—but private groups are trying to fill in the gap.

For example, the World Space Foundation (a private group of space engineers and enthusiasts in Los Angeles) is designing and actually fabricating a small light sail for deployment and operation in space, if it can hitch a ride on an available rocket. In Europe, the British Interplanetary Society is considering a "space sailing regatta" for a race to the moon, in which several competing groups of designers (possibly financed by advertising written on the sails), would, by remote control, jockey their orbiting sails to maximize solar thrust and spiral outwards from Earth. The winner would be the first team whose sail passed behind the moon as viewed from London. Such giant sails would be easily visible to ground observers with binoculars, and the months-long race would be likely to arouse tremendous world-wide enthusiasm.

Rob Staehle of the World Space Foundation reported to the

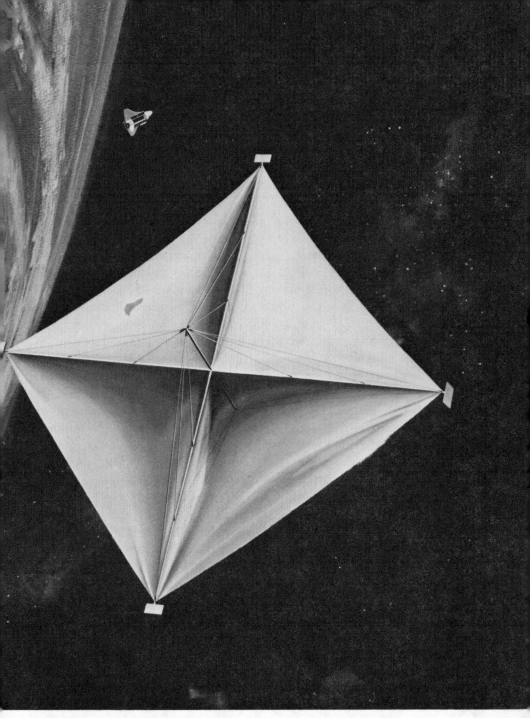

Light sails offer a unique as well as economical source of solar system transportation.

Boulder conference on a much larger design suitable for supporting a Mars expedition. This WSF design called for a square metallic sail more than a mile on a side, weighing about nineteen tons including structural support. The unmanned remote-controlled vehicle could tow thirty tons of supplies from a staging area in low orbit near Earth all the way to Mars orbit in less than two years. That is three times as long as the standard coasting trajectory normally proposed for use by the manned vehicle. But the time should not matter because nobody would be eating or breathing aboard the supply boat.

The official Case for Mars Conference summary took note of the long future development schedule for sail propulsion, compared with the hoped-for rapid advent of manned Mars flights. "If developed fully," the report stated, "the solar sail may be the propulsion system of choice for travel in the inner solar system. In the time frames of this Mars Exploration Program, however, it seems unlikely that solar sails can be utilized as manned transportation. . . . However, for unmanned cargo and smaller payloads the solar sail concept offers a highly viable resupply option and a cargo (or sample) return vehicle which can be used in a more or less continuous fashion to ferry cargo and samples back and forth between Earth and Mars."

Electric propulsion uses a fundamentally different means of expelling propellant to obtain reaction force. With chemical engines it is the heat of the ferocious chemical combustion which ejects the gases outwards; with nuclear systems, pure heat causes violent expansion of any fluid, but the lighter it is, the better, so it will reach the highest possible ejection velocities. With electric systems, either electrostatic repulsion of like-charged ions and engine surfaces (ion-drive), *or* electric currents shoving a stream of hot gas (plasma drive) create extremely high expulsion velocities. The "specific impulse" efficiency of an engine is directly related to this expulsion velocity, and electric engines offer values at least ten times as high as the best chemical systems, but at a very low thrust. Nevertheless, since they can fire for months at a time, they can build up to extremely high velocities.

The concept of electric propulsion has been around since the mid-1950s, and small-scale technological work has been going on for decades. In 1970, a NASA satellite called SERT was put into

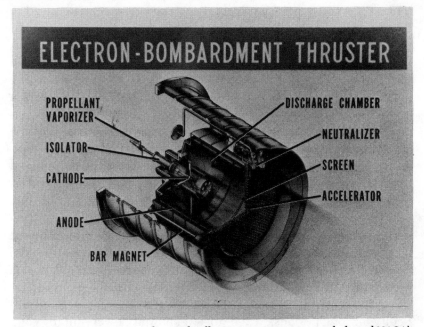

ELECTRON-BOMBARDMENT THRUSTER

PROPELLANT VAPORIZER

ISOLATOR

CATHODE

ANODE

BAR MAGNET

DISCHARGE CHAMBER

NEUTRALIZER

SCREEN

ACCELERATOR

This ion engine was operated periodically over a ten-year interval aboard NASA's SERT-II (Space Electric Rocket Test) mission. Much larger engines could take manned or robot cargo ships to Mars and back.

orbit carrying ion thrusters which then were test-fired over a ten-year period; the Soviets had flight-tested a small ion-drive engine using argon propellant aboard Yantar space vehicles in 1966. Although these tests worked, no follow-on efforts were made. In 1976, John Newbauer, a professional space technology magazine editor, quipped, "Electric propulsion has always been the propulsion system of tomorrow; will it ever become the propulsion system of today?" Five years later, his magazine, *Astronautics and Aeronautics*, published a progress report. "Severe emotional peaks and valleys characterize the electric-propulsion evolutionary cycle, sometimes testing the durability of even the most dedicated of its supporters. . . . The end of 1981 again finds the electric-propulsion effort in this country without a designated primary mission and with little indication that there will be a NASA-sponsored program in the near future to develop one." Despite its revolutionary potential, electric propulsion technology remains essentially dead in the water.

Early concepts saw the electric engine drawing its prodigious power requirements from atomic reactors; these vehicles had small sail-like structures which were actually radiators to dump waste heat from the nuclear power plants. By the mid-1960s, engineering difficulties with long-life nuclear systems, combined with evolutionary trends toward cheap, lightweight, highly efficient solar cell arrays, saw a general transition to "*Solar* Electric Propulsion System" (SEPS)—in which giant banks of solar power panels created the necessary electrical power to expel the propellant and thus create thrust.

The best SEPS propellants are either metallic conductors (such as mercury) or ionizable fluids (such as argon gas). One such recent design, from the NASA Lewis Research Center in Cleveland (where NASA electric propulsion work is centralized), extended known thruster technology up in scale to a three-foot-diameter thruster, using argon (liquid argon can be stored cryogenically without difficulty for long periods), a 200- to 400-kilowatt power unit (by comparison, Skylab and the Space Shuttle run on less than twenty kilowatts), and a specific impulse of about 10,000 seconds. Such a design is considered easily buildable—whenever the need arises.

Such power systems for electric engines have applications even when the spacecraft is resting at Mars between its interplanetary voyaging (turning it off might cause more damage than just leaving it idle). The electrical power could be beamed down to receivers at the surface base, or be used at one of the small moons of Mars—primarily to process local material for the manufacture of air, water, and even propellant. Other uses for a powerful electrical power generator at Mars could also be imagined (imaging radar to probe the surface and subsurface for liquid water, for example).

Because of the low thrust of SEPS and sails, vehicles powered by them would take many months to spiral outwards from parking orbit near Earth to the point where they could break free and begin their interplanetary trajectory. Careful analysis has shown that the optimum combination of propulsion schemes might be a high-thrust system to achieve escape velocity, and then the high-efficiency system for the interplanetary long-haul. If the low-thrust systems are to be used for the whole escape energy, they incur "kinematic inefficiencies" that require twice the cumulative delta-V as that which would be enough for a high-thrust direct departure trajectory; half is "wasted" on the long creeping spiral climb out of Earth's gravity well.

But the advantage still lies with such systems, if perfected, since they have excess efficiency to spare. They do *not* produce faster transfer times to other planets than do high-thrust systems (since they take so long in building up their top speeds), but since they only use one-tenth as much propellant their starting weights are a lot lower. This in turn leads to considerable savings in launch and construction costs.

One additional problem with such low-thrust systems was identified twenty years ago by space propulsion expert Maxwell Hunter. "It has been found by experience in rocket propulsion," he wrote, "that static tests should be run which sum to an equivalent of perhaps 100 times the propulsion period of the vehicle in question. . . . Clearly, operating a [low-thrust] engine for a period of time equivalent to 100 trips is impractical. . . . But just as clearly, from what we know of the statistics of development, something like this would be required to insure reliable operation. . . . When the relative difficulty of developing reliable hardware is considered, high-thrust vehicles have considerable advantage."

If these SEPS and sail techniques are not sufficiently mature or sufficiently well tested to be "man-rated" by the turn of the century, they can still support such manned expeditions by transporting cargo to emplace supply caches near Mars. Such storable supplies and equipment can be sent ahead of the fast manned craft by "slowboat."

Careful loading of the unmanned supply fleet is crucial to mission success, so that the loss of one ship would not jeopardize the whole mission. In fact, two types of cargo might be identified: mission-critical items, such as propellant or mining gear (things astronauts *must* have), and mission-enhancement items, such as drills, extra rovers, shelters, and lab equipment (things astronauts *would like* to have). Loss of the former could not be tolerated; loss of some of the latter items would allow the mission to proceed with less ambitious intentions.

Several of the "delta-V maneuvers" (speed changes) on the Mars voyage involve slowing down instead of speeding up. Arrival at Mars is one such case—you have to shed at least 10,000 fps or you'll go sailing right on past the planet.

One extremely attractive technique is to use the air drag of the target planet's own atmosphere. The alternative is to use forward-facing rocket thrusting alone to carry out the slowing down. This entails a great deal of propellant, while the "aerocapture" maneuver may be possible with the addition of a special "aeroshell" shield plus some maneuvering systems to steer the atmosphere-grazing flight path to the precision required.

"Aerocapture" has been defined as the maneuver in which a passing spacecraft sheds enough of its velocity to enter an orbit around the target planet. It is an extreme case of "aerobraking," which merely refers to any slowing of a space vehicle achieved via passes through the upper layers of a planet's atmosphere—whether the probe is captured or not. Both applications are like steering a runaway car (with its brakes failed) into a row of hedges to bring it gently to a manageable speed.

One NASA study for an unmanned Mars orbiter highlighted the potential value of aerocapture. Starting with a given spacecraft weight on the approach to Mars, a vehicle using aerocapture could put more than twice as much payload weight into Mars orbit as could a vehicle using rocket engines alone. The numbers were 78 percent and 36 percent of the total pre-maneuver weight, respectively. Three-fifths of the propulsive vehicle had to consist of propellant to be burned; one-fifth of the aerocapture vehicle had to consist of the aeroshell to protect and steer through the aerodynamic stresses.

One special cost of aerocapture must be pointed out. It can be quite stressful to the crew, with up to six to eight Gs of deceleration over a period of about a minute. That stress can be somewhat reduced by the addition of structural modifications of the aeroshell to produce more lift, but any additional weight reduces the advantage of aerocapture over propulsive braking, so it must be carefully balanced.

Crewmembers will have to endure this brief but violent load after eight to ten months of weightlessness. Some space doctors feel there are no medical dangers as long as the astronauts are lying down and are not expected to move during or soon after the high-G period. Others feel that such a load would be unacceptably high and could cause bone fractures. The value in spacecraft weight savings must be balanced off against the medical dangers.

In conclusion, the different mission phases of the Mars expedition can be compared to the menu of available or potential propulsion techniques. Some of these follow.

Ascent from Earth to orbit. The Space Shuttles of the 1990s would be sufficient but would need to devote a large fraction of a year's total flights to carrying the pieces up. A reasonable development such as a Shuttle-derived cargo vehicle would be extremely useful since it would allow the Mars spaceship to be assembled with only "several" launches.

Injection on the trajectory to Mars. Liquid-fuelled systems using hydrogen are entirely capable of doing the initial missions; nuclear systems would be desirable but are not worth the costs of building them from scratch. For unmanned supply ships, either light sails ("solar sails") or electrical propulsion may prove extremely attractive.

Braking into orbit around Mars. Aerocapture is the desired option, unless medical problems with the high-G deceleration prove unavoidable. Propulsive braking is feasible, but it increases the weight of the space vehicle two or three times.

Descent to and return from the martian surface. The kind of space vehicle described in an earlier chapter, a combination of the Apollo-style Command Module shape for entry and the Apollo Lunar Module design for ascent, is the best of the known options. No propulsion breakthroughs or shortcuts are foreseen here.

Injection on the return trajectory to Earth. The same systems that supported the original departure from Earth would serve here. One option is to build the propulsion system so that it can be refuelled at or near Mars from local material processed during the long layover.

Braking at Earth arrival. Aerocapture is preferred here as well, but some homeward-bound trajectories have speeds far too high for heat-shielded capsules to endure, so at least partial propulsive braking may be required.

Now, there are "wild cards" waiting in the wings. Serious people speak confidently of breakthroughs in gas-core nuclear fission, or fusion engines (the "rigatron"), or pulsed fusion systems, but they cannot realistically be expected to be available for the first man-to-Mars expeditions at the turn of the century. But the lesson of this chapter is that such breakthroughs are not needed. We already have the propulsion technology to carry us to Mars—in fact, we have had it for many years.

*When ships to sail the void between the stars have been built,
there will step forth men to sail these ships.*

Johannes Kepler

3

Human Factors

IF MOST OF the great exploratory expeditions of the past—those
of Columbus, Magellan, Cook, Lewis & Clark, Scott & Amund-
sen—had been required to meet the safety and comfort standards
expected today from space expeditions, they never could have been
made. Their ship losses and personnel losses were substantial,
even—on tragic occasions—total.

That could not be allowed to happen on the first expeditions
to Mars, whatever the cost. For reasons based firmly on the scope
of national commitment to such programs, the man-to-Mars effort
must guarantee that the crew will stay alive the whole trip, without
untoward medical or psychological complications, and perform their
duties as well as the hardware allows them to.

Such a requirement is practically asking for a miracle, but such
a miracle is not too much to expect from modern "space medicine."

Mars flight provides four main threats to human life. First,
basic life support involves adequate diets and avoidance of buildups
of contaminants or other toxic substances. Second, there is the
problem of physical adaptation to spaceflight conditions, and the

65

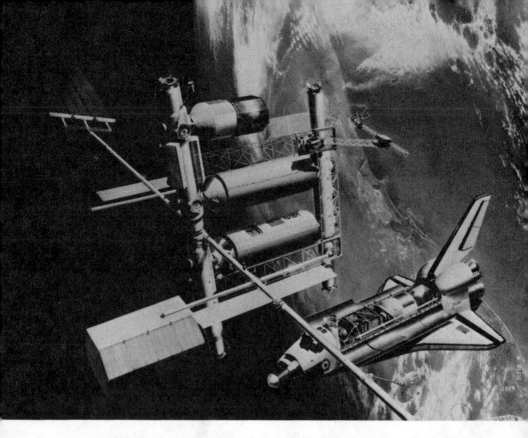

Space stations of the 1990s will provide crucial medical, life-support, and operational experience relevant to manned flight to Mars.

consequent danger of returning a too-well-adapted human body to a gravity environment. Third is the unique environmental danger of outer space, particularly radiation. And last are the traditional problems associated with exploration (such as traumatic injury) and isolation (such as possible degradation of the immune system, or psychological problems leading to carelessness and costly mistakes). Each of these threats can be counteracted or endured, but the cost is high.

The physical requirements for keeping a human body alive are straightforward. Supply enough oxygen (several pounds per day per crewmember) and remove exhaled carbon dioxide. Supply water (about four pounds a day) and nutritive, digestible food (about three pounds, two-thirds of which is water); dispose of the water given off by crewmembers as sweat or exhalation on one hand, and as urine on the other (in about equal amounts), and dispose of the dry fecal material.

In practice, a life-support system can be "open," in which all oxygen and water is provided fresh, or "closed," in which waste products are processed to provide recycled oxygen and water, at a considerable cost in equipment weight and power. The open system is simple but can be costly on long flights, since the weight at launch is directly proportional to the duration of the flight. The closed system can use heavy, complex recycling equipment, but costs far less in weight for every additional day of the mission.

Also in practice, the degree of recycling does not have to be high: the first 50 percent can be relatively easy (the Soviets have been doing it for years, recovering exhaled water from cabin humidity), the next 30 percent can be difficult, the next 15 percent

REGENERATIVE vs OPEN LOOP EC/LSS

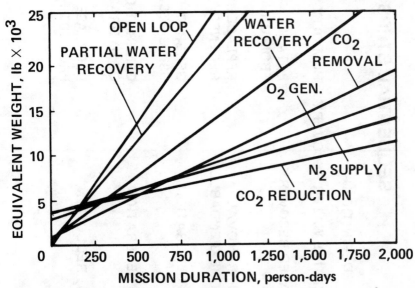

A closed-loop life-support system is almost essential for such long interplanetary voyages. This chart shows how much long-term weight is saved by increasing the initial weight of regenerative equipment.

SEQUENTIAL STEPS IN LOOP CLOSURE

DEFINITION	DESCRIPTION
PARTIAL WATER RECOVERY	HUMIDITY CONDENSATE COLLECTION
WATER RECOVERY	POTABLE WATER RECOVERY AND TREATMENT FROM URINE AND WASH WATER
CO_2 REMOVAL (NONEXPENDABLE)	REPLACEMENT OF EXPENDABLE LiOH WITH REGENERATIVE CO_2 COLLECTION TECHNIQUE
O_2 GENERATION	GENERATION OF O_2 THROUGH WATER ELECTROLYSIS USING RECLAIMED WATER
N_2 GENERATION	GENERATION OF N_2 THROUGH DISSOCIATION OF HYDRAZINE
CO_2 REDUCTION	DECREASE IN EXPENDABLE WATER BY RECOVERING PRODUCT FROM CO_2 REDUCTION (SABATIER) PROCESS

can be extremely costly, and the final few percent, unattainable in practice. But one rarely needs to go that far. "A little bit of cheating goes a long way," remarked biologist Bassett Maguire at Boulder, discussing the desirability of fully regenerative systems. A strategy that recycles with 70 to 80 percent efficiency would be more than adequate for a Mars expedition; the extra cost in equipment weight to push the percentage up much higher just would not pay.

Dr. Philip Quattrone of the NASA Ames Research Center, one of the world's leading experts on space life-support systems, told the Boulder conference that in-space testing of recycling systems was an absolute *must* for the long Mars expeditions. Plans for a late-1980s American permanent space station called the SOC (Space Operations Center) include life-support technology for carbon dioxide recycling. "We need SOC," Quattrone told the conference. "If SOC doesn't go, we're not going to Mars." Given the go-ahead on SOC, he believed it would take about a decade to have a first-generation system in place aboard the space station, where it would accumulate a few years of experience. That in turn would lead to another decade's development of a second-generation

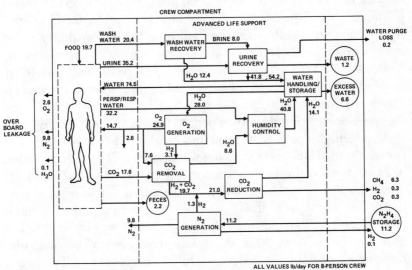

Life-support systems for space stations will be a proving ground for the equipment to be used on Mars flights.

high-efficiency system suitable for a Mars expedition. Quattrone did not see that point arriving until 2005 to 2010.

The problem of water recovery from urine seems paramount. Relatively simple "space stills" are not much help. A complex series of pumps, compressors, filters, treatment chemicals, and storage tanks is required, and I've tasted the best products possible using 1981-era technology: there's a long way to go before it's drinkable.

For a turn-of-the-century man-to-Mars expedition, a life-support system using some chemical and physical recycling schemes was deemed sufficient by Quattrone and other Boulder conferees, even if it were far from the most sophisticated systems which would be coming available soon afterwards. Quattrone also suggested that prototype biology-based systems might be available to assist in air regeneration and purification, but they were not essential. The human beings on the voyage would thus be provided with the air, water, and food they needed to stay alive, and the use of recycling could save thousands of pounds of spacecraft weight on even the shortest possible expeditions (that amounts to several percent of total payload, so it would pay to develop such technology as soon as possible).

During the months of flight towards Mars, the bodies of the crewmembers will come to feel at home in the condition of weightlessness. This adaptive physiological change is not in itself harmful, but it has a great effect on the health and work capability when the crewmember is returned abruptly to a weighty environment. Under the best of conditions, movement will be difficult and tiring; under the worst, the crewmember could suffer serious or even fatal physical injuries.

At Boulder, Dr. Daniel Woodard (a young Houston physician and spaceflight enthusiast who has worked at the NASA space center there) described the changes that occur. "The redistribution of fluids toward the head, experienced at the very start, initiates a complicated series of events," he reported. "The head senses that it has an excess of fluids" and the crewmembers thirst stimuli are reduced. "Blood plasma volume is lowered by 11 percent. The number of red blood cells, along with their oxygen-carrying component, hemoglobin, is reduced a full 11 percent. . . . The heart, which is pumping less blood and is not pumping against gravity, shrinks a little and, in the words of a NASA doctor, 'gets lazy.' The

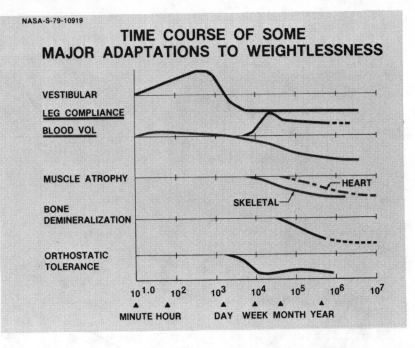

NASA-S-79-10919

TIME COURSE OF SOME MAJOR ADAPTATIONS TO WEIGHTLESSNESS

VESTIBULAR

LEG COMPLIANCE

BLOOD VOL

MUSCLE ATROPHY — HEART

SKELETAL

BONE DEMINERALIZATION

ORTHOSTATIC TOLERANCE

$10^{1.0}$ 10^2 10^3 10^4 10^5 10^6 10^7

MINUTE HOUR DAY WEEK MONTH YEAR

leg muscles, which have almost no work to do, atrophy and lose their strength."

Exercise is one answer to maintain heart and muscle conditioning. Soviet spacemen in six-month-long expeditions worked for two or three hours a day on different exercise programs, which came to an equivalent of climbing a 200-story building each day. Better conceived exercise programs might take as little as an hour or two every other day, but this must be confirmed in flights aboard space stations over the next ten to fifteen years.

A more insidious adaptation process is the so-called bone decalcification, or bone demineralization. Up to half a percent of bone calcium is lost every month, and a 10 percent loss is considered pretty serious. Soviet data suggest that the loss rate declines after a few months, but that has not been confirmed. Devising countermeasures is stymied by the unfortunate fact that there is no general agreement on what really causes this effect in the first place.

One Russian (Valeriy Ryumin) has made two back-to-back, half-year flights without any detectable physical damage.

"There are several theories," Woodard explained at Boulder, "ranging from the malabsorption of calcium in the intestinal tract to effects of fluid shifts on the parathyroid gland, to electrolyte disturbances and vitamin deficiencies [also proposed has been the effect of rapid movement through Earth's magnetic field]. But most evidence suggests that bone demineralization is simply a normal adaptive process of the weight-bearing parts of the skeleton [but less marked on other bones] in an environment in which they are no longer subjected to stress. . . . Because it's not a disease, there may be no 'cure'. . . ."

Exercise and special diets have not shown any effect in space, but some experts believe it was the wrong kind of exercise and the wrong kind of diet, and new flight experiments will be performed in coming years. Some drugs, notably the diphosphonates, seem to slow down calcium loss in bed-rest patients on Earth (strictly horizontal bedrest is a good analog for spaceflight weightlessness). Ultimately, the bone weakening may just have to be endured, but it's hard to believe that the next decade or two of manned space activities will *not* find effective countermeasures to this, the most insidious physical effect of spaceflight.

Radiation is a fact of life in space beyond the protective shield of our planetary magnetosphere. Woodard called it "a major source of concern in a flight to Mars," and most specialists agree. In the least, sufficient amounts of radiation (measured in rems) would be absorbed by Mars explorers to probably terminate their active spaceflight careers, along with statistically significant additional risks of developing cancer years after returning to Earth (a vigorous life-long diagnostic process could probably prevent a rise in death rate attributable to these extra flight-induced cancers, primarily leukemia). In the worst cases, rare but ferocious solar storms could sicken or even kill individuals or the entire crew.

The rem count determines the health impact. Twenty or thirty rems is virtually harmless, reported Woodard, while 300 rems can kill, and 1,000 rems will definitely do so. Current federal health standards allow a total of 300 rems exposure over an entire career (an average of about ten rems per year).

Interplanetary voyages will accumulate about 40 rems per year from cosmic rays and background solar radiation. Unpredictable solar flares can deliver upwards of 60 rems in an average year, with bad years producing several times that amount. For protection against such short, fierce solar flares, the Mars-bound astronauts will need to have a "storm cellar" in the ship: this could be at least a small area designed to place as much mass as possible between the crewmembers and the sun; at best, it could include special radiation-absorbing, high-density shields or bulkheads. Vulnerability to such radiation bursts continues in Mars orbit and to a slightly lesser extent on the martian surface, since Mars has no protective magnetic field like Earth's; surface explorers might need to build their own shelter under a roof of several feet of dirt.

"What kinds of medical problems can we expect to encounter on a flight to Mars?" asked Woodard. Previous space mission medical kits were designed to accommodate typical human maladies. "Coughs and congestions, chest pains, bowel and digestive problems, gastro-intestinal infections, urinary infections, joint inflammation, headaches, vision disturbances, toothaches, cuts and bruises, fractures, skin infections, and so on," Woodard explained. "On a long mission to Mars, it might be necessary to perform surgery with sterile fields, to administer anesthesia and to diagnose with X rays. Items which will probably have to be included on a Mars mission are blood chemistry and hematology capabilities, dental equipment, medical manuals, respiratory support, surgery equipment, EKG monitors and all manner of drugs and medications, anesthesia, intravenous fluid equipment and so on."

Basic healing processes may be changed by weightlessness. Although incisions, bruises, and similar flesh wounds may not be affected, some space doctors (notably astronaut William Fisher) have expressed concern that bone fractures may *not* heal properly without significant gravitational stresses.

Woodard added it all up: "If one looks at all the necessary equipment for a good health maintenance facility—from exercise equipment like a treadmill and a bicycle ergometer to all the diagnostic and surgical equipment necessary to meet most emergencies—one is looking at approximately 1,000 pounds of equipment. Naturally, the more complete and elaborate the medical facilities, the higher the cost; the less complete and elaborate the medical facilities, the greater the risk to the crew." Because of the critical danger in this one area of life support, Woodard recommended (and most other studies concur) that one member of the crew be a physician with practical experience in emergency medicine; at least two other crewmembers should receive extensive backup medical assistance training. Coming from a physician with precisely the kind of experience being specified, Woodard's criteria might seem a little self-serving, but since most of the people at the conference had given up any hope of getting to Mars themselves, nobody objected to the young doctor's enthusiasm. And his professional judgment was sound: medical care of the crew on the years-long voyage requires the ultimate in house calls, a live-in doctor.

The first few days on the martian surface will present a unique medical problem. In the past, spacemen returning to Earth from long orbital voyages have been able to relax and undergo careful tending by teams of specialists in well-equipped hostels. Not so on Mars—the crewmembers will be on their own, and be faced with the need for energetic physical activity from the moment their craft enters the martian atmosphere.

Soviet experience shows that this requirement can be met, with careful preparation. Exercise during the outbound leg is essential; special high-salt and high-fluid diets just prior to landing (to boost the depleted blood volume) have also proved useful, as have *intravenous* fluids. After six months in space, Soviet cosmonauts have been able to walk to waiting helicopters, and they are strolling through the park outside their hotel by the third day back, all this in a full-G environment. Mars is less than 0.4 G, and while the outbound leg will be somewhat longer than the six months endured by cosmonauts in 1979 and 1980, more sophisticated physical training methods and equipment will no doubt be developed in the next twenty years.

Besides, the first week's surface activity could be specially tailored to present a reduced workload. Early walks on the surface, to set up equipment and grab initial samples, could be made on long umbilical hoses rather than with heavy backpacks. The major portions of early EVA sorties could be scheduled to be in a seated position in the Mars jeeps which could be deployed as early as the first day.

One particular danger involves the necessary readaptation of the body's balance mechanism and the retraining of reflexes learned in a weightless environment. Skylab astronauts reported that their vestibular functions remained "short circuited" for several days after landing and that they could orient their bodies only by sight, not by any sense of balance. They also reported an alarming tendency to drop things, since they had become accustomed to letting go of tools, packages, or other objects in midair and of unthinkingly assuming that they would stay put until grabbed again, which on Earth, to the surprise of the astronauts, they consistently refused to do! Both problems will occur during the early days of the Mars landing, and the crew must simply be alert to the possibilities of serious mistakes.

When I first mentioned to my friend "BJ" Bluth, a leading space psychologist who attended the Boulder conference, that I would be able to spend only a few pages on the psychological problems of man-to-Mars expeditions, she told me I should have my head examined. Books could be written on the topic, and, as the mission nears, no doubt books *will* be written (one of them, I hope, will be by her). But in this book, many crucial issues have already been abbreviated so as to present an overview of the "big picture," and the psychological bottom line is that with careful preparation (much *more* care than has been exhibited so far in the American space program) the psychological stresses involved in such a long and tension-filled mission *can* be accommodated.

At Boulder, Billie Jean Bluth discussed the terrestrial analogs of Mars expeditions in terms of psychological stresses. She singled out Navy submarine duty as being most similar, with regard to hierarchies, mission duration, isolation, physical danger, level of comfort, and other factors. "On submarine crews we can study patterns of depression and activities to dispel depression. In particular, we can observe communications breakdowns and displaced hostility—which in space would be directed against the people in Mission Control." Psychological stress can have physical manifestations which fellow crewmembers and Mission Control experts might be able to notice and diagnose: drowsiness, disorganization, misinterpretation, hostility, a seeking for changes (often destructive), daydreaming and absentmindedness. Soviet psychologists monitor the stress level of cosmonauts by doing a "voice stress analysis" on the harmonics of the crewmen's voices, a technique applicable on the Mars mission.

"The Soviets have a lot to teach us on the psychological support of long duration missions," Bluth pointed out, "because they've had all the recent experience." A special support team attached to the Moscow Mission Control Flight Surgeon's office specializes in psychological support of the orbiting cosmonauts. They constantly plan ways to introduce novelty into the crew's routines, via "surprise packages" sent up from Earth inside supply ships, plus conversations both with family, with friends, and with total strangers who have interesting professions or backgrounds. Even before flight, Soviet space psychologists work not as antagonists of the cosmonauts but as their allies. Rather than use psychological investigations to screen out all candidates without natural psychological strength, the Soviets actively teach their cosmonauts how to cope with expected psychological stresses of space missions. Such an approach, advised Bluth, is the way American space doctors will have to learn to function for the Mars mission.

"There should be an odd number of crewmembers," Bluth also urged, "so as to avoid deadlocks. Experience has shown that

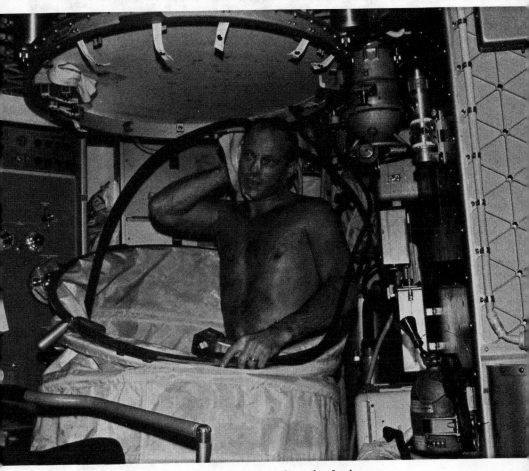

Showers provide psychological as well as physiological refreshment.

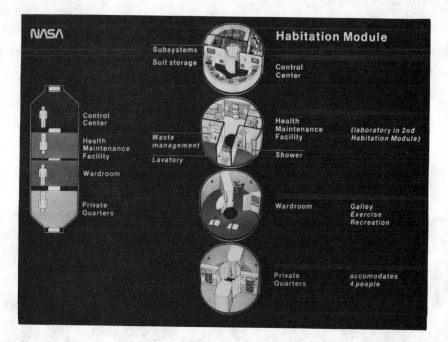

Two different architectural approaches to space station sections are also applicable to the design of the Mission Module living quarters where the Mars astronauts will live for two years.

even numbers of people under stress tend more often to split into two equal and opposing camps, unable to reach a democratic solution to urgent mission decisions."

All of these health-related factors come together to specify the characteristics of a "best" astronaut crewmember for the Mars expedition. The flight might be the last space mission for all crewmembers, due to cumulative physiological damage. People experienced under long space expeditions, probably aboard space station facilities, would be preferred. Maturity and an absence of separation stresses involved with minor dependent children might call for older personnel.

These desired characteristics might tend to force the selection of people between the ages of forty-five and fifty-five years old, not the popular image of a dashing young "jet jockey" that the term *astronaut* brings to mind. But a crew with an average age of about fifty would probably be the most capable possible selection.

Besides, if the mission occurs at the turn of the century, a few astronauts who will be about fifty years old by then are the ones who are in their early thirties now, and there are several of them even now in the astronaut program who have their eyes on Mars already. And at least two of them are doctors!

4

The Outbound Leg

THE ROAD FROM Earth to Mars and back contains no milestones
or signposts, and its maps are represented with abstruse mathematical jargon about "gravity fields of influence," "delta-V," and
"hyperbolic excess velocity." But celestial navigation, even under
unearthly rules, is still humanly conceivable. The routes were being
mapped out in theory long before they were attainable in practice.
The question facing analysts now is not which is the way to go, but
rather what factors of propulsion and the duration of the mission
are most important in choosing between dozens of alternate routes.

The paths of objects orbiting the sun depend on their speed,
direction of motion, and distance from the sun and from any nearby
planets. Mathematicians love to solve the resulting equations and
analyze them for aesthetic elegance; computers are perfect tools
for crunching out precise navigational numbers. But most of us are
just interested in getting there and back.

The cheapest path between the planets is the "Hohmann
Ellipse" which swings the spaceship halfway around the sun on the

journey of about 260 days. In practice, because the orbital planes of Mars and Earth are not precisely parallel, the best trajectories travel either a bit under half an orbit ("Type I") or a bit over ("Type II") and hence can range in duration from 220 to 300 days depending on the particular launch geometry of the planets.

Because the "Hohmann Ellipse" requires Mars to be at the one point where the ellipse kisses the orbit of the target planet Mars, one and only one relative alignment of Earth and Mars will allow a spaceship to make the trip and have Mars there when *it*

SA-S-70-6112 X

The outbound leg begins with the firing of powerful rocket stages which have been assembled in a low parking orbit near Earth.

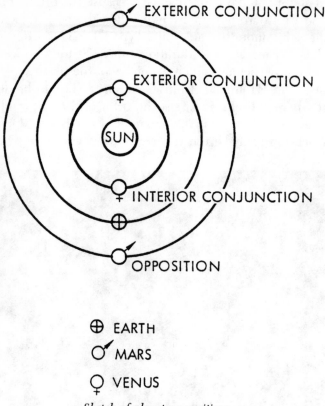

EXTERIOR CONJUNCTION

EXTERIOR CONJUNCTION

SUN

INTERIOR CONJUNCTION

OPPOSITION

⊕ EARTH
♂ MARS
♀ VENUS

Sketch of planetary positions.

gets there. This relative position defines a "launch window" and occurs when Mars is about forty-four degrees ahead of Earth, a relationship which repeats itself every twenty-five to twenty-six months. Because of the ellipticity of the orbit of Mars, the amount of energy required to make the transfer varies significantly, from about 50 percent to 60 percent over and above the velocity already needed to get into orbit around Earth. To get back to Earth from orbit around Mars, another velocity change is required of about three-quarters as much.

Because of the changing positions of Earth and Mars, the minimum launch window back to Earth occurs 460 days after the arrival of the outbound ship. The total mission time, 260 plus 460 plus 260, or 980 days, is probably too long for an initial expedition.

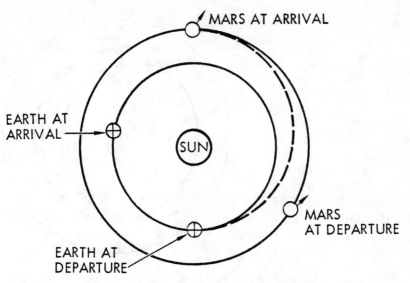

The simple "Hohmann ellipse" approach to Mars.

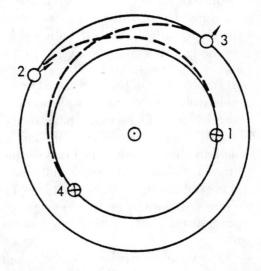

1. EARTH DEPARTURE
2. MARS ARRIVAL
3. MARS DEPARTURE
4. EARTH ARRIVAL

The minimum-energy "conjunction-class" mission requires a very long stayover at Mars before heading back to Earth.

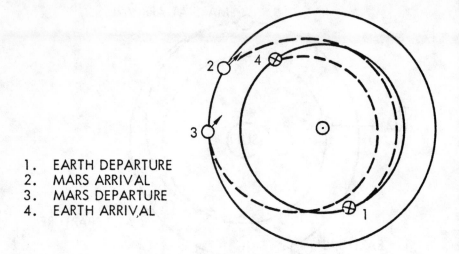

1. EARTH DEPARTURE
2. MARS ARRIVAL
3. MARS DEPARTURE
4. EARTH ARRIV,AL

A more costly (in terms of rocket power) profile involves staying only briefly at Mars before making a near-sun swoop to catch up with Earth. It's called an "opposition-class" mission, and its advantage is a mission of shorter duration.

For an extra amount of propellant, the travel time can be cut by one-third by staying at Mars for only a few weeks and then returning on a trajectory that dives in close to the sun and allows the ship to catch up with Earth on the other side of its trajectory. If Venus is properly positioned, the spaceship can use its gravity to further improve the return path. These Venus fly-by missions offer the best combination of relatively short mission times, 600 to 700 days, at relatively little cost in additional propellant.

Such long flight times have many implications for mission planning. The outbound and inbound legs will involve low levels of activity, but the crew must be kept busy for health reasons. The journey is so long that a significant amount of time (at least weeks, and possibly many months) must be spent at the destination. Total mission durations of up to two years will place great demands on spacecraft reliability, propulsion systems storability, and life support.

The common wisdom about the long voyage up and out from Earth to Mars is that boredom and restlessness will be a major concern. A well-known space reporter wrote in *Spaceflight* in 1971 that "Boredom and zero gravity will be the crew's biggest enemies.

Boredom, which will exist only until planetfall, will be alleviated by making extensive tape libraries available to the crew." Out of the spaceship's windows, no visible sign of the vehicle's headlong progress (at more than 100,000 feet per second) would be visible, and the only movement on the celestial sphere would be the week-by-week crawl of the brighter planets, a motion compounded by the spaceship's own orbital track and the changing parallax of shifting positions.

The most noticeable and distressing symptom of the widening separation from Earth would be the growing round-trip radio communication lag, a delay which would mark the advent of an isolation hitherto unknown to space voyagers, as human conversation with colleagues and family members back earth-side became impossible. Such psychological pressures, gnawing at supposedly idle minds drifting between the worlds, could well seriously diminish the mental health and alertness of the voyagers—at least, that's what numerous commentators have alleged.

Far from it. If recent spaceflight experience is any evidence, the Mars-bound crew will be overworked and constantly challenged. Far from brooding over the view of an unchanging field of stars, the voyagers might not have time to look out the window for

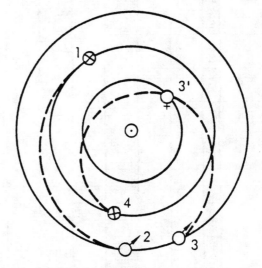

1. EARTH DEPARTURE
2. MARS ARRIVAL
3. MARS DEPARTURE
3' VENUS PASSAGE
4. EARTH ARRIVAL

This "near-sun swoop" can be improved still further if Venus is positioned near the flight path, so that its gravity can be used to modify the spacecraft's flight.

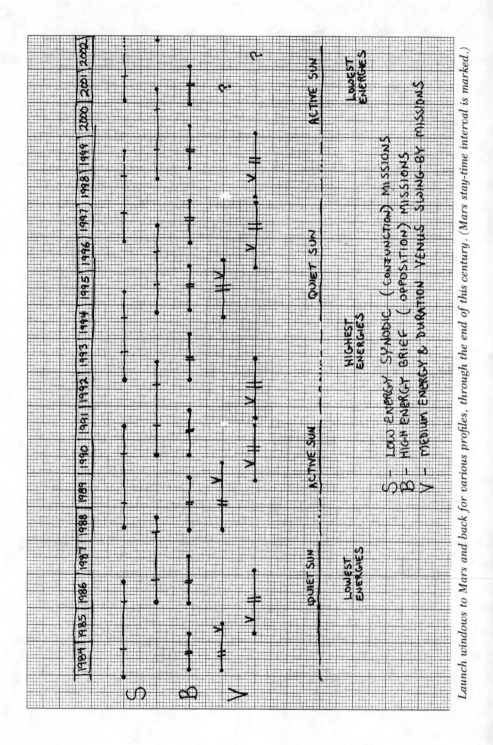

Launch windows to Mars and back for various profiles, through the end of this century. (Mars stay-time interval is marked.)

days on end. Far from sitting by the radio hungry for voices from Earth, the astronauts will probably come to see radio communications as an unwanted interruption of their busy schedules. There will be plenty of things to do in those eight to ten months en route.

The first order of business on the way outwards, once the whole spacecraft has been thoroughly checked out and all systems are exercised, will be to prepare for an immediate return to Earth, but only to be carried out in the event of a major spacecraft or crew emergency. For the first several months, as the spaceship drifts slightly ahead of and ever further out from Earth, the crew has the ability to use onboard propulsion to turn back and get safely to Earth within a few weeks at most. But after about a third of the way out to Mars, the delta-V required for this abort maneuver has grown to exceed the capabilities of their vehicle's engines, and the expedition is committed to at least a Mars fly-by and a long loop around the sun.

Other mission abort modes can be formulated, and the outbound crew will spend a lot of time studying them, practicing them, and carrying out the preparations needed to keep such options open on the shortest notice. For example, a decision to cancel the entire Mars orbital/landing phase of the mission would require a major course change at Mars fly-by (probably using the Mars Entry Module's descent and ascent propulsion systems, unless that system's breakdown was the original cause of the wave-off) and again halfway back to Earth. But the necessary emergency return trajectories can be computed and would already be loaded into the navigation system pre-mission.

In the most dangerous abort case, failure of the ignition of the MEM's ascent stage on Mars would strand the crewmen on the surface, but it need not be fatal if the waiting mother ship (the "Mission Module") can drop its life-support supplies (and any astronauts still aboard, orbiting Mars) to the surface, possibly inside what otherwise would have been the crew's Earth Entry Module. The whole stranded expedition can then sit tight and go on reduced rations to await a pickup by a rescue ship launched from Earth at the approaching next window. (Additionally, if needed, a high energy, unmanned, resupply flight could pay a delta-V penalty to allow an off-nominal launch and get a six- or seven-month flight time, carrying several tons of supplies, enough to allow the crew to last through the long wait.

Surprisingly, on the long synodic (1,000-day) missions the schedule is not particularly strained. A pre-existing spacecraft (probably the one originally slated for the next major expedition, and hence already in an advanced state of preparation) could be reloaded in the four months between the realization of the MEM malfunction and the already-scheduled second expedition's departure from Earth. The rescue ship would then follow a nominal trajectory and arrive at Mars only three months later than the stranded astronauts would have gotten back to Earth anyhow. They would thus have to stretch about 260 days of supplies (the return leg) to last them 350 days (allowing for the wait until the arrival of the rescue expedition)—and that's three-quarters rations, which in a pinch could be quite manageable for an inactive, extremely cautious and patient crew of stranded Mars astronauts.

But that's only if things go wrong. Even when everything is running smoothly on the outbound leg, there are several distinct activities to fill the too-few days between Earth and Mars. Keeping alive and keeping the spaceship operating would certainly be significant activity, with frequent checkout runs and diagnostic inspections, but added to all of that would be the major tasks of *study* and *practice* for the Mars-side mission events. Scheduling lessons derived from all previous manned space programs will be needed to fit everything that needs to be done into the scant time available.

Probably the single most important factor in the success of the American manned space program over the decades has been the extent of crew control built into the spacecraft, and the consequently required exhaustive training of the flight crews and ground crews (at Mission Control, Houston, and elsewhere). On numerous occasions, hardware problems have been overcome by flexibility, ingenuity, and on-site alteration of preplanned sequences. Life-threatening failures have been successfully finessed by well-trained, alert personnel in space and on Earth.

To reach this level of expertise, crewmembers work long hours on training mockups, practicing and continuously improving the procedures which have been written for nominal and emergency developments. The most sophisticated training is in the spacecraft simulators, which are hooked up to powerful computers that read all crew commands, then calculate what should be the consequence of such commands were the spacecraft really in flight. Finally the computers drive the cockpit displays (including artificially produced views out of the windows) to show what the real effects would have made them show.

The heart of this process of training is the "integrated simulation," or "sim" for short. A crew sits in their simulator, while the simulation computers feed their data into Mission Control computers, just as the data from the real spacecraft would be fed. Dozens of specialists ("flight controllers," in a specialized hierarchy under the "flight director" who has ultimate authority in the mission) monitor the health and happiness of the computer's imaginary spacecraft. A third group of training specialists observe (but are not observed by) the crew and the flight controllers, and deliberately introduce certain malfunctions into the simulator computer's concept of the state of the imaginary spacecraft. These malfunctions then affect the data output to the crew and to Mission Control, who must in turn react to the failures, diagnose them quickly and accurately, and repair them, negate them, accommodate them, decide to ignore them (false indications are an authentic class of hardware failure), or—if there is no other choice— abort the mission soon enough to save the lives of the crew (too cautious, and you might be tricked into aborting a repairable mission; too bold, and the astronauts "die" before they can reach safety).

These training specialists have a crucial and unsung role in the success of actual missions, even beyond their contribution to honing the participants to a razor's edge of sharpness. They also seek to find the most subtle and damaging failures or combinations of failures, and must therefore be first-rate engineering systems analysts in their own right, with a touch of the sadist, the conjurer, and the clairvoyant thrown in. Vulnerabilities which they uncover in the make-believe world of "sims" (and it's a learning process which continues right up until the actual space mission) are repaired in reality by changes in hardware, software, or procedures.

Such activity has worked for manned spaceflight in the past, and worked very well. The Mars mission will need this very same type of service, but it will have to be long-distance.

All of the exquisite training computers and staff now located at the Johnson Space Center in Houston will still be available, but the minutes-long radio round-trip time from Earth to the expeditionary spacecraft will add a new dimension to the complexity of the crew training problem.

Most definitely, high fidelity training will be needed on the flight. The crucial portions of the mission, Mars orbit insertion (call it "MOI") and the landing itself, will take place almost a year after the last chance for earth-side training. There's no way the crewmen can store away those reflexes and learned procedures to remain dormant in their minds for such a period. At the very least, refresher training must be scheduled, but if so, the outbound voyage then might as well include the major portion of the key training for the at-Mars activities anyway.

Therefore, the Mars spacecraft must be designed from the beginning with in-flight training in mind. All flight controls (buttons, switches, keyboards, control sticks, etc.) must be able to operate either in direct ("real-life") mode or in simulation ("make-believe") mode. Similarly, all flight displays (gauges, television screens, lights, etc.) must be controllable in either of those modes, too. With the spacecraft in simulator mode, another computer system somewhere else on board must maintain the mental mathematical model of the "imaginary" practice spacecraft with its hypothetical practice problems.

This dual-mode operation scheme, while certainly an innovation for manned spaceflight, is feasible because of an advance in spaceship flight control systems characterized by the Space Shuttle's on-board computer quintet. The technical term is "fly-by-wire," a technique used as a backup system in earlier manned spacecraft which has become the primary and only control system on the Space Shuttle. Essentially, all controller commands—engine firings, elevon pitch, gauge readings, whatever—come from the computer system, based on measurements taken throughout the vehicle, including from the crew's flight control switches, sticks, and buttons. Such a scheme, in which the computer system is programmed to select different control combinations depending on rapidly changing mission phases, was the only one judged capable of handling the intricate requirements of the Space Shuttle mission—but it required, in turn, a major advance in control theory and reliability of airborne computer systems.

So day by day, the Mars-bound astronauts would undergo landing simulations. Some could be quite normal, familiarizing them with the actual steps they hopefully would be following for the actual touchdown. Other runs would include difficulties which had to be detected, recognized, and circumvented—all with the aid of advice from Mission Control millions of miles away (where the data flow would be artificially delayed on tape to match the expected real delay connected with the distances at the time of the real landing). Since the simulator control of the Mars spaceship could not tolerate any such delay, another on-board computer would have to act as the simulator control, thus requiring a high level of capability and special software programs. Perhaps the lander's computer system could double as the simulator computer for the mission module's simulations of Mars orbit insertion; in turn, the mission module's computers could double as the simulator computer when it came time for the lander (the Mars Entry Module, or MEM) to practice its own particular specialties.

Such simulations, involving the whole crew (one would act as an on-board training official, while all the rest would be in actual training), could probably be scheduled as often as three times a week, for eight to ten hours each time. That heavy load is in fact characteristic of astronaut training for major new missions, and was followed by both the Apollo-11 moon-landing crew in the six months before their 1969 mission and by the first Space Shuttle Columbia crew over the same general time period before their 1981 mission. And it was far from the hardest part of their preparation.

In fact, drawing on training experience for such analogous astronaut activities, the three full days of simulations per week would be just part of the crammed crew activities. They would also undergo special procedural and equipment briefings, probably for three half-days a week, and be assigned to other specialized instruction on individual duties (probably in the form of videotapes or even motion holographs) two other half-days. Their testing of the spacecraft for routine diagnostic functions would likely consume a couple of hours per crewman, twice a week, and general space housekeeping, judging from Skylab space station experience, would require about two hours per day per astronaut (such duties as communications sessions, navigation updates, corrections to on-board documentation, and so forth).

Also from Skylab experience, the rest of each astronaut's day can be fairly well mapped out: sleeping and personal hygiene, nine to ten hours per day; food preparation and mealtimes, two hours per day (and maybe three on Sundays); exercise to prevent the deterioration of muscles to be needed for walking around on Mars, at least two hours a day on average (maybe with Sunday off); major medical examination, one or two hours per crewmember per week; personal time, an occasional hour here or there, and a big block set aside on Sunday; but if Skylab is any example, that last item, together with sleeping time, is the most easily sacrificed when it comes to actually fitting everything together.

Somewhere in there, too, must be fit extensive scientific training for the Mars phase of the mission. Lander crewmembers would be expected to earn the equivalent of a correspondence course master's degree (at least) in geology; orbiter Mission Module crewmen, if any, would be doing the same but preparing primarily for visual and instrumental observations from orbit. Each crewmember would also have collateral backup training in specialties of other crewmembers, in the event of somebody's being incapacitated. The Mars spaceship would take on all of the spirit and appearance of a flying university library in the week before final exams!

Bored crewmembers on the way out to Mars, you say? Don't they wish!

The interruptions of communications from Earth may not be very welcome in such an environment. One particular aspect of such communications, however, may make it more convenient: it will not occur in "real-time," since at the very least there will be five- or ten-minute delays induced by the great distances involved. Conversations in the ordinary sense will probably become impossible about 1 or 2 million miles out, when the round-trip delay becomes an appreciable fraction of the minute. Beyond that point, the astronauts and Mission Control will probably talk to each other in "message packets" of conversation, carefully enumerating the kinds of responses expected back at the other end of the very long distance line. In fact, a handy technique to create the illusion of conversation, and to jog the memory of the astronauts about what the Mission Control spokesman is talking about, would be for Houston to rebroadcast to the spaceship a taped replay of their own preceding snippet of conversation followed immediately by

Houston's response to it. The crewmembers could also later review selected portions of such tapes.

Time measurement of another sort could be a source of confusion on the flight out. There is one unique feature of the martian environment which will eventually require a physical adjustment on the part of the crew, and that is its length of day. The Earth-Mars journey is probably the best place to make the switchover so any possible stresses will be eased by the time the astronauts face the other stresses of the actual landings.

Perhaps there will be very little stress, since the martian day is so similar to that of Earth: from sunrise to sunrise on Mars averages twenty-four hours, thirty-nine and one-half minutes. The extra forty minutes results in what was called a "sol" during the Viking mission, a specialized space term for "one Mars day." Sometime soon after leaving Earth, the astronauts (and also shifts at Mission Control) will probably want to switch to "sols" rather than "days."

So all Mars-mode timepieces will thus have to squeeze an extra forty minutes into their cycles. There are various subtle and sophisticated ways of doing it (one suggestion is to "stretch" each ordinary minute by three-fifths of a second), but the simplest way is probably best: tag it on after midnight, as forty "leap minutes." At midnight, the digital clocks would read 23:59, then 24:00, then 24:01, up to 24:39 and some seconds, then leap to the next minute, 00:00, then 00:01, and a new day would have begun (it's safe to say that mechanical wristwatches could present something of a problem!). For psychological reasons, the Mars clock hours should run close to local landing-site sun time, with noon near local solar midday.

At Mission Control and aboard the spaceship there would be several sets of clocks (or, for economy, one clock with several operating modes). Current Mars time would be displayed back in Houston, but of much greater operational utility would be two biased displays: one would show the Mars time at which data and voice signals currently arriving at Mission Control actually left the spaceship (this would allow flight controllers to coordinate such signals with past schedules in the crew activity plan); the other would display the spaceship time at which voice and data sent up immediately from Houston would actually reach the spaceship (this would allow flight controllers to coordinate messages with actual future scheduled mission activities). There's no question that such complications would take a lot of getting used to, in space and on Earth.

I leave as an exercise to the reader the development of a Mars-side *calendar* to be used aboard the spaceship.

Seventy-three days out, the crew would have a chance to observe an extraordinary sight: the transit of Earth and moon across the face of the sun, as they pulled ahead of the slowing spaceship trajectory. In centuries past, observations of such solar transits of Venus and Mercury were crucial in providing data on the scale and schedule of the solar system, but such data is no longer of particular scientific urgency. Psychologically, however, the eight-hour-long event could be an important one, marking one of the few demonstrable milestones on their measureless path between planets.

Other astronomical observations may be more valuable, but would have to depend on the unique position of the spacecraft, a position not attainable by larger observatories back on Earth. Possibly foremost among these observations would be those of *occultations*, when rocky bodies in the solar system pass in front of distant stars. The flickering of the starlight can tell astronomers much about both the nature of the occulting object and about the nature of the distant star (is it, for example, a very close binary?); the rings of Uranus and possible satellites of some asteroids are among recent valuable discoveries made by observing occultations.

The spaceship's path, in a slightly different plane from Earth's, would open up an entirely new set of occultations which would never have been visible from Earth. These can involve asteroids, outer planets and their satellites, or even Earth's moon. Both real science, and real crew psychology, would be served by such a program of tracking two or three occultations (each takes only minutes) a week on the journey.

As the months drag by, the busy Mars-bound astronauts are certain to appreciate any possible distraction from their work loads. One likely source of comic relief would be the discovery of a mythical mascot for the expedition, an alien entity already said to lurk the trans-Mars celestial highways ready to gobble up, or at least grab a bite out of, passing spacecraft. That creature is known among spaceflight operations personnel as "the Great Galactic Ghoul," and it (he? she?) is tailor-made to appeal to the practical jokester in a Mars explorer-astronaut.

The legend of the ghoul is authentic, albeit tongue-in-cheek. It dates from 1969, when a Mars-bound Mariner space probe encountered a small "glitch" shortly before reaching the planet. In a drinking party attended by flight controllers and newsmen, the idea came up that the failure could not have been a coincidence, since some earlier probes had also experienced problems at that point on their approach trajectory, and two Russian probes had even gone dead. The string of coincidences was woven into a Bermuda Triangle of space, wherein dwelled some malevolent, hungry monster which swallowed Russian spacecraft, but only took bites out of American ones ("It's probably the glue," one engineer speculated). An artist later created a delightful painting of the mythical beast, fat and ugly and weirdly colored, floating in space picking its teeth as a well-chewed Mariner space probe limped away. And in later years, as other Mariner and Viking and Soviet probes approached the final leg of their journey, flight controllers made jocular references to the upcoming death duel with the Ghoul, a duel which, to the relief of everyone involved, the probes all survived.

The opportunity for comic relief is too great to be missed, and by that point—almost eight months and more than 200 million long, coasting miles out from Earth—the crew will deserve every laugh and temporary relaxation they can muster. Mars lies just ahead.

5

Site Selection

ALMOST ANY POINT on Mars would teach us a great deal about
the planet. And some points are certain to be safer, or richer in
resources useful to explorers than other points. But no one point
can teach all the lessons. Only one point can be first, so where
would it be?

Engineers and scientists are bound to come into conflict over
this point, and would be replaying for Mars the same arguments
fought over Apollo and the moon. "The problem," one NASA space
geologist summarized, "is that the safe places are the boring places."
That is, the locations which the engineers, mission controllers, and
astronauts prefer also happen to be in general the least interesting
scientifically. (The Viking landing sites were disparagingly called
"the blandlands" by scientists eager for more interesting locations.)

So what would be the scientific community's input to the
choice of early landing sites? Admittedly, it will not be the only
input and may not even be the dominant input early on, since

safety and possibly surface resources will rate highly. But from scientific rationale, a philosophical issue must first be settled: to go for generalities or specifics.

The "Bogard Report," a 1977 discussion of unmanned sample return mission goals, worded it this way:

> Site selection based on characterization of specific surface features (e.g., lobate ejecta patterns) rather than on solution of basic planetological problems (e.g., the origin of martian volcanic rocks) seems to be argued by some investigators. However, such a strategy generally cannot answer most of the planet-wide questions. . . . All of the lunar landing missions were oriented toward planet-wide problems, rather than placing high priority on the elucidation of specific surface features, such as the lunar sinuous rilles. Even Apollo-15, landing adjacent to the Hadley Rille, was primarily directed towards returning a diversity of samples which could provide information on broad internal and external processes of the

Many sites on Mars, such as the volcano Uranus pictured here, allow access to highly varied terrain without much travel.

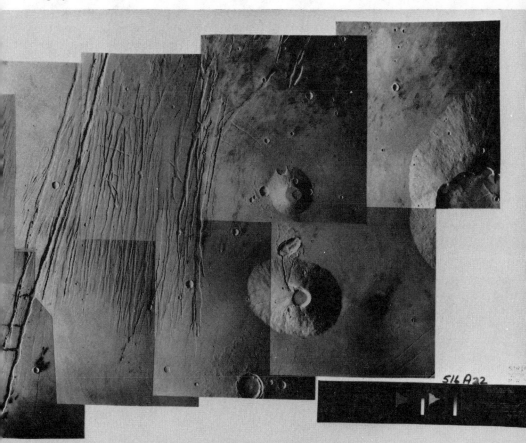

moon; evidence pertinent to the origin of the rille was a lower ranked objective . . .

With this philosophy in mind, the Houston space geologists listed five types of potential sample collection sites and the mission scientific objectives which would be pursued within ten to fifteen miles of such landing points. The five site classifications were:

1. The bottom of the Coprates Canyon (Vallis Marineris), near contact with cratered plains and ancient cratered terrain beyond the rim of the canyon at a landslide or talus slope, could answer questions on the absolute age of old plains volcanism, the age of ancient cratered terrain, the depositional processes on the canyon floor, weathering throughout Mars' history, the history of volatiles including water, general atmospheric studies, along with mechanisms of canyon formation, landslide mechanics, and dune material. (On Earth, you would look for that kind of information in the Grand Canyon or the Olduvai Gorge, Kenya.)

2. The flank of some young volcanic feature, near the boundary of ancient cratered plains, could answer questions in geochemistry, the absolute age of recent volcanism, weathering processes (including wind transport), atmospheric studies, plus the absolute ages of volcanic plains and superimposed impact craters, along with the age of adjacent volcanic and impact features. (An Earth analogy would be fresh volcanoes in Sicily or Hawaii.)

3. Sites on some volcanic plains filling a major uplands basin, near mountains, could answer questions about the age of volcanics filling the basin, the depositional processes and products in intermountain basins, plus general geochemical, petrological, and atmospheric studies, along with the geochemistry of ancient cratered uplands, the age of basin formation and of meteorite impact features. (On Earth, this would be like the Northwestern United States, or the Deccan Plateau in India.)

4. A site on the ancient cratered highlands, near the base of an ancient impact crater cut by gullies, could answer questions on the age of crustal materials, erosional and depositional processes, and the presence of water. (The Laurentian Shield in Canada, or a similar structure in Scandanavia is an earth-side analogy.)

5. An ancient river delta would be an ideal site to answer questions about paleontology (if the subject matter, "life," ever

existed) along with the history of sedimentary processes and of water-transported material brought from great distances upstream (think of the Nile or Yangtze Deltas).

Other planetary scientists have developed different, even more detailed strategies. At JPL, home of the Viking probes, follow-on studies for an automated sample return mission were conducted as late as 1980. One contractor group was the Planetary Science Institute, headed by Dr. James Cutts (who came to Boulder).

Three particular areas were identified by the Cutts Report as providing multiple types of accessible terrain. "The sampling ranges required for reaching a significant set of objectives seldom exceed a few kilometers," the report claimed, "because of the occurrence of sharp contact relationships between different rock types on Mars." First was an area of "grooved terrain" near volcanoes and ancient cratered terrain, located east of the majestic Olympus Mons caldera; second was a major erosional feature called the Kasei Vallis, dotted with young volcanoes; third was Olympus Mons itself, where its lava flows encroach on the much older "grooved terrain." Also nominated for early sampling missions was some unspecified site in the polar regions, near the edge of the permanent ice cap.

To get even more specific, I asked Mars geologist Hal Masursky (of the U.S. Geological Service office in Flagstaff, Arizona) what his choice was, since he had been scheduled to deliver a paper at Boulder on that very topic. Without hesitation, Masursky responded, "I'd go for the Big Valley."

The giant Valles Marineris, spanning thousands of miles along the martian equator, is an impressive feature even viewed from far out in space. The Grand Canyon of Arizona could be tucked away unnoticed in one of the martian valley's side tributaries. Like the Grand Canyon, but on a scale several times as deep, the martian valley cuts through layer upon layer of ancient material (lava, wind-blown dust, ejecta, and just possibly water-borne sediments) which are exposed to the surface for collection and analysis.

Two promising valley sites have already been identified by planetary scientists: the Candor Chasma ("Shining Canyon") and the Hebes Chasma. A 1980 JPL study characterized them. "Candor and Hebes canyons are two branches of Valles Marineris situated in the northwestern part of the Coprates quadrangle of Mars. These and other canyons in the area are thought to be dropped fault troughs (grabens) which were extensively modified by later eolian [wind] activity. . . . A large volume of material has been removed from the canyons. . . . The canyon caprock material is composed of basaltic flows that form the eastern part of the high Tharsis plateau. . . . The ancient, heavily cratered terrain, which appears at the surface nearly 2,000 kilometers east of the canyons, is covered in the site by a thick sequence of volcanic flows and interbedded pyroclastic deposits. . . ."

Two groups of rock layers seem to be exposed on the sloping canyon walls: the upper sequence is composed of resistant, thick layers (possibly lava flows interbedded with volcanic ejecta); the lower sequence is more thinly layered and evidently was eroded layer by layer during deposition. Ancient crustal rocks may be exposed on the canyon bottom (which is also covered by extensive dune fields), while very young lava flows cover the surrounding plateau surface.

To reach outcrops of representative samples covering more than four miles of vertical profile, a surface traverse of more than 120 miles will be necessary. This is due to various exposures and blockages caused by landslides, dunes, and other modifications of the vertical sequence. Lava samples collected from each layer can be age-dated via radioactive-dating techniques, thus establishing the absolute age of the whole sequence. Soils mixed into the layers, thus dated, can time-tag planetwide channel formation, and possibly show evidence of ancient climate variability. Weathered zones between dated layers can be correlated with the ages of water erosion episodes and volcanic eruptions elsewhere on Mars. A whole calendar of the planet's evolution could thus be unrolled from this one site.

The traverse (or actually a series of traverses) would be difficult but feasible. Valley slopes are not too precipitous in most places, although many landslides against canyon walls testify to ground instability. The canyon floor has a great abundance of rocks and boulders and could provide isolated rough spots which would have to be circumvented. Considering the local topography, the best landing zone is atop the smooth, dust-covered Candor Mesa right in the middle of this section of the canyon. It could provide a perfect base camp for sorties in all directions.

A formal landing site selection study sponsored by JPL in 1980,

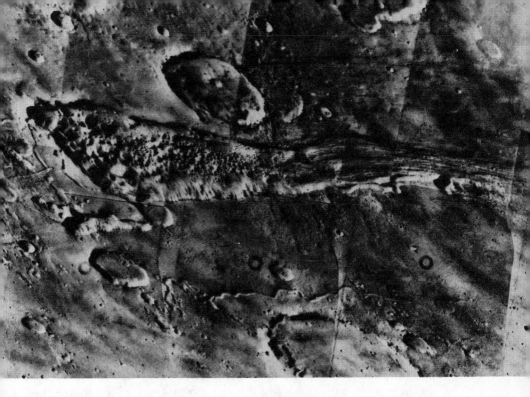

This Viking orbiter view of Capri illustrates the massive volume of water which once erupted here from beneath the permafrost and flowed downhill to the right.

and headed by Hal Masursky, summarized the argument for Candor this way: "It offers a unique opportunity to sample rock layers and their interbedded soils that would reveal the petrochemical history, age dates, and the environmental changes that may correlate with episodes of channel formation and the history of solar variations preserved in the rocks from the time when they were exposed at the surface." For that reason, the "Big Valley" is obviously science's number one choice when it comes to the first human footsteps on Mars.

The ancient history of Mars (including the general question of solar system origin and planetary formation) could be sought at another site located in the ancient, cratered uplands of the southern hemisphere. The strategy here would be to seek a fresh crater which had penetrated more recent lava flows and hence excavated ancient crustal material.

Masursky's team looked at this selection, too, and finally settled on a spot called Tyrrhena Terra. It had the requisite deep craters (the recent lava flows there are about half a mile thick) along with windblown dust deposits and an old complex canyon system a mile wide, possibly containing gravel from distant upstream sites. It is not far from Tyrrhena Patera, identified by another geology team as one of the oldest martian central vent volcanoes, of which only four have been found (they are a remnant of an unusual style of pyroclastic volcanism that occurred early in the planet's history and then tapered off).

Another candidate ancient site, comparable in age to the lunar highlands explored by Apollo-16, is a hilly cratered region southeast of the Schiaparelli Basin. Geologists believe it abounds with breccias and crustal rocks formed during the high bombardment era in solar system history, long before Earth's surface first solidified.

(Because of their relatively smooth surfaces and near-equator locations, these ancient highland sites might become leading candidates for the first manned landing. The other sites—valley, polar, volcano, etc.—require much more precise touchdown and longer staytimes for most production science and resource exploitation, so they are excellent possibilities for expeditions which follow the first proof-of-concept surface mission.)

Recent volcanic sites are also of interest. The Houston "Bogard Report" explained their significance this way. "Interpretations of the absolute age of the youngest volcanic features range from a few hundred million years to 2.5 billion years. This range of interpretations is equivalent to the differences between an Earth-like history of continuing volcanism and a moon-like history of very early volcanism with an ancient cutoff of internal activity." Thus, these sites could provide crucial data pertaining to the evolution of the whole planet.

Under contract for the 1979–1980 JPL study on *unmanned* sample return candidate sites, a team of geologists from Arizona State, headed by Dr. Ronald Greeley, examined Viking and Mariner orbital photography for candidate young volcanoes. Youth could be judged only in relative terms, as the Bogard Report had cautioned, but was ascertained mainly by seeking regions with the fewest impact craters. The requirements also included areas with nearby safe, smooth-landing zones, of low elevation (for air braking), and preferably near the equator.

Greeley's team examined sites on the Tharsis Plateau (not

young enough), Alba Patera (too hazardous), the Elysium volcanoes (too uncertain), and Arsia Mons—finally selecting a site in the western reaches of that complex. The site is characterized by multiple lava flows, often hundreds of miles long, dozens of miles wide, having edges 150 feet high with forty-five-degree slopes; the lava is derived from vents on or near Arsia Mons itself. General slopes in the area are under ten degrees; the volcano, towering more than 50,000 feet above the surrounding plain, has the largest caldera of any of the shield volcanoes (it's a hundred miles across), with a structurally fascinating rim displaying multiple terracing just waiting for a geologist's pick hammer. Exactly where such features fit into martian geologic history remains undetermined, and the age-dating of rock specimens is the only currently known way to find out.

The most recent phases of martian history are recorded in the polar regions, where erosion and layering processes seem to be active right up to the present day—since craters are extremely rare on the surfaces near the north and south polar caps.

The objectives of a polar mission would contrast sharply with those at other sites. The prime emphasis would be stratigraphy—the analysis of layers of material which could act as pages in the book of planetary history.

Layers more than a hundred feet thick, piled atop one another, can be seen in photography from orbiting spacecraft. Each of these is probably broken down into smaller layers, possibly only as thick as sheets of paper. Both deposition and erosion have been active near the polar cap and there are deep gorges where thick sequences of layers can be reached without the need to drill.

A great deal can be determined from analysis of the dust trapped in these layers. Variations in grain texture and size can specify the air pressure under which the sand was abraded and laid down. Long-term variations in temperature can be determined by analysis of the variation of the relative concentration of stable isotopes of oxygen and carbon, although useful amounts of carbon-14, so handy for dating historical sites on Earth, should not be present on Mars in any abundance due to the low nitrogen content of the atmosphere (terrestrial carbon-14 is made from nitrogen by cosmic ray interactions). The purposes of such measurements would be to chart the recent climatic record of Mars, with its indications both of solar energy output variability and of other internal processes which affect a planet's (any planet, including Earth's) climate.

The Cutts Report pinpointed one accessible site, called "Site B," at 84.5° N, 105° W, close to different types of interesting terrain (including gullies) yet in a smooth, apparently safe landing zone. And not only is it an obviously valuable site for science, it is one of the few regions where even now we can be certain there is access to water.

Another object of search would be organic materials trapped and preserved within the ice; no organic material at all was found by the Vikings, *but* there should have been detectable amounts of such material deposited by meteorites, so some surface process is destroying those compounds. Also, as in Antarctica on Earth, valuable meteorites should be found on the ice caps, perhaps of types rare or nonexistent on Earth.

The poles of Mars may not always have been where they are now. To search for really ancient sediments dating back more than a billion years or two, explorers will have to seek out where the old poles (the "paleo-poles") have wandered to, and two scientists recently thought they had found just such features near the modern equator.

Peter H. Schultz and Anne B. Lutz-Garihan of the Lunar and Planetary Institute in Houston suggested in 1980 that three equatorial areas with up to a mile of layered sediment may have been in polar regions before the great Tharsis volcanoes built up a lava bulge so thick as to imbalance the whole planet's spin, and cause the rotation axis to shift. Valleys in these areas show spiral patterns similar to those forming now around the modern poles.

The largest and best preserved region lies near the equator near 155° W, and is probably at least as old as the Arsia Mons volcano. A smaller, more eroded region is near 180° W, and by counting craters, the scientists have estimated it is about as old as the Tharsis lava plains. The smallest and most heavily stripped region, near 210° W, probably predates the volcanic action which formed the Tharsis uplands. Shifts between these sites and the current north pole probably occurred during various events simultaneous with the Olympus Mons volcanic eruptions.

If so, these sites could preserve extremely ancient climatic records of the sun's behavior more than a billion years ago, the period when life on Earth was still confined to the oceans.

Not all volcanic activity on Mars is ancient. Some may be recent, and some may even have been spotted by the Viking Orbiters. Outside of the scientific value of recent or still-active volcanic activity, such zones may be areas of outgassing of useful volatile materials including water, and *hot* water at that. Iceland-style hot water heating systems for future Mars bases become a possibility—and where there is warm water with lots of energy sources nearby, there too is a candidate environment for life forms, either native martians or visiting aliens from Earth.

In 1981 at the Third Mars Colloquium, scientists turned their attention to evidence for a very recent large-scale explosive event on the Hecates Tholus volcano (32° N, 209° W), possibly a martian Mount St. Helens. West of the caldera, orbital photographs showed very fresh-looking ground, probably thick deposits of air-borne volcanic ash. "Recent" could have been up to a million years ago, or it could have been a century ago, or it could have been a day before the photos were taken.

Leonard Martin of the Planetary Research Center, Lowell Observatory, in Flagstaff, claimed at the same colloquium that "the Viking Orbiter cameras recorded several events that may be of volcanic origin." On August 30, 1977, an unusual small dense cloud extended upwards from the surface within the south rift zone of Arsia Mons volcano. The cloud was centered directly over a small crater chain in three sets of photographs taken a half hour apart. Martin pointed out that the cloud and the shadow seem to meet at the surface, where the cloud is brightest and the shadow darkest. The conclusion—that a portion of Arsia Mons is still erupting gently after a billion years or more—was not widely accepted at the conference, but it gained additional corroboration a few months later when geologists at a Venus conference published results suggesting that massive active volcanism still persists on the surface of Venus, at geological features which may be similar to the Tharsis uplands on Mars.

A small unnamed mountain north of Solis Lacus, at latitude 16° S, longitude 80° W, was also tagged by Martin as evidencing potential volcanic activity. Alternately, he suggested, it may be a possible geyser or steam vent, based on clouds seen in Viking photographs.

One martian surface feature which achieved particular popular notoriety in 1980 is the "Great Stone Face." It certainly is a bizarre-looking structure in the single photograph taken of it, like a face staring into the sky. There were widely published suggestions that the "face" and some nearby pyramidal mountains were *not* natural formations.

"Unusual Martian Surface Features" is the title of a pamphlet written by Vincent DiPietro and Greg Molenaar in 1980, which described their computer enhancement of the Viking Orbiter pho-

The Arsia Mons volcano is an ideal site for the first expedition which plans to "winter over" at Mars, staying the full 400 days needed for a conjunction-class (synodic) mission.

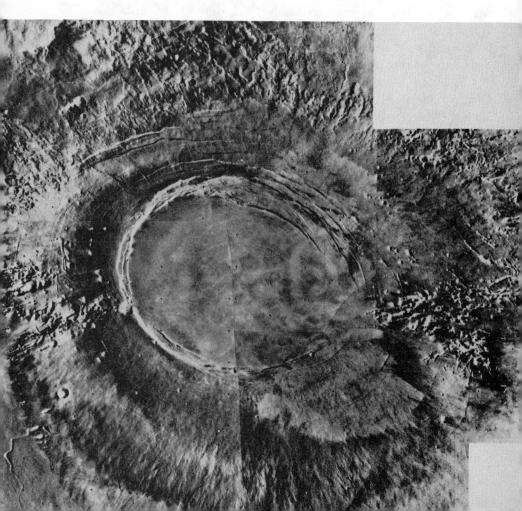

This pair of Viking orbiter photos shows what might be a geyser of volcanic eruption near Arsia Mons.

tograph showing the mile-wide feature located at 41.0° N, 9.4° W. "The writers assume no conclusions with evidence of the surface features found and their resemblances," the authors temporized cautiously. "The writers do assert that these features are very unusual and that further investigation be performed in this area." DiPietro and Molenaar, both computer specialists associated with the NASA Goddard Space Center in Greenbelt, Maryland, came to the Boulder conference and contributed substantially to study groups on advanced surface reconnaissance—just the kind of techniques that could resolve the mystery of the "stone face"—if there is a mystery, of course.

Pyramidal natural structures are widespread on Earth's own deserts, for example, and there is a remarkable "stone face" tourist site in Vermont which testifies to the variety of naturally occurring shapes stone can achieve accidentally. Space geologist and desert

The "stone face" is a fascinating structure, whatever its origin. New generations of orbiting probes will probably solve its riddle without need of a landing—unless it actually does turn out to be artificial—highly unlikely, but. . . .

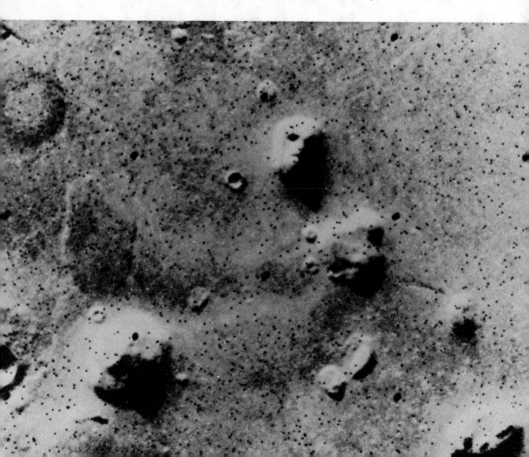

expert Dr. Farouk El-Baz, formerly the astronauts' own private tutor on lunar geology, wrote about such features in the *Smithsonian Magazine* in 1981. He explained how natural pyramid shapes, which he found while exploring the Egyptian deserts west of the Kharga Oasis and near the Gilf Kebir plateau, resist being eroded away completely by the abrasive sand-bearing winds. "The conical shape evades destruction by leading the wind upslope and funneling its erosive power to the peak where its energy dissipates in the air." El Baz also described natural features called *yardangs* which look a great deal like unfinished statues of the sphinx. The features on the martian "great stone face" are almost certainly similar: a natural plateau with a few weirdly scattered hills on top, which under one (and only one) sun angle cast shadows which give the illusion of eyebrows, nose, and mouth.

But even in that case, the erosion processes in the area are bound to be of scientific interest.

Consider now this different set of requirements for a Mars landing location. It was compiled at the Boulder conference by the "Mission Strategy Workshop" chaired by James French of JPL.

> The criteria for site selection will be of extreme importance. A first cut at what these criteria might be was addressed by the Workshop. . . . The first and most obvious is that of accessibility. We must be able to conveniently reach the surface site chosen for the base from reasonably expectable arrival trajectories. . . . Next in importance is the presence of the necessary raw material. This must be absolutely certain prior to the commitment to a landing. . . . A third very desirable characteristic is low elevation of the base site simply because this allows us to take much better advantage of the martian atmosphere in final deceleration. . . . Another factor of substantial importance is our ability to communicate with Earth. It is desirable to have a site which will be in contact, that is to say line of sight, through the martian year for at least part of the day. . . . Of medium importance would be the ready availability of solar energy which would tend to dictate a lower latitude site. . . .

At the very end of the list, French included one additional factor. "Since presumably various means of transportation . . . will be available to the crew to visit sites of scientific interest, it is not of maximum importance that the base site itself be a highly scientifically interesting site."

The resources of Mars are extensive, compared to nearly every other accessible object in the solar system; there will be an entire chapter soon on how to "mine" air, water, and even food and construction materials on Mars. But of all the valuable resources needed to support manned expeditions there, *water* heads the list in value and in required amounts.

WATER LOCATION

Where would you go to look for water on Mars? One easy and nontrivial answer is in the polar caps, and since there are good scientific reasons for landing there anyway from time to time, the search for water will be an easy one. But most other interesting sites are a long way from the poles, so finding ready sources of water on or below the surface could be a challenge.

There apparently *used* to be lots of water flowing across the martian surface, leaving traces in the intricate networks of gullies and channels. But that was probably about four billion years ago, and even then the water didn't stay long. There is no evidence, for example, that there ever was any standing water (oceans, lakes, or even ponds) anywhere on the surface, since orbital photography shows nothing looking like beaches. Nor is there unambiguous evidence that it has *ever* rained from the martian sky. Water seems to have sprung from localized regions, perhaps as artesian wells breaking through the crust (occasional floods of that type seemed to have occurred throughout Mars' history), while drainage-like branching valley networks now are generally interpreted as not representing rainwater runoff patterns but rather the uphill "sapping" (or undermining) of melting subsurface ice. That process also could be continuing at a slow rate even today.

Most Mars geologists believe that permafrost (a mixture of frozen soil and water) within tens of feet of the surface is widespread over much of the planet. The uniquely martian shapes of numerous surface features seem to suggest this interpretation: craters with muddy-looking ejecta blankets, seen nowhere else in the solar system; chaotic terrain looking like collapsed surfaces after the

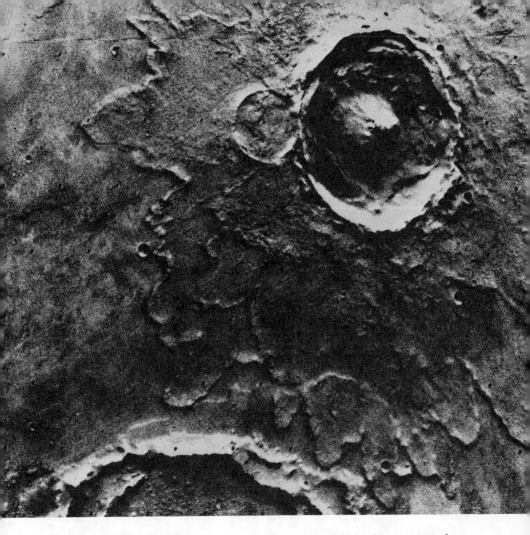

Muddy-looking craters are now generally interpreted to imply the existence of soggy soil, and the water is probably still there.

melting of buried ice sheets; volcanic features similar to those found in Iceland where lava pours out under, above, or through glaciers.

Since the planet has a source of internal heat that once was strong enough to push massive amounts of lava up through the crust, much of that heat must remain, insuring that the hypothetical permafrost layer cannot extend very deep. One Soviet analysis puts the depth at about half a mile near the equator and up to two miles in the polar regions. Below that depth, water could exist as a liquid in fractures and porous gravel, until impermeable bedrock is encountered about five or ten miles down.

In some specific nonpolar regions, the water ice may be closer to the surface and hence more accessible. These regions would be pinpointed by the proposed "Prospector" Mars polar-orbiting "water-sniffer"; a seismic network would also help to detect subsurface ice *if* there were enough naturally occurring quakes to "sound out" the nearby structures. But even now there are serious suggestions for two types of water-enriched surface areas: "swamps" such as near Solis Lacus, and "glaciers" in the Tharsis region. If either feature really exists, exploration/exploitation strategies will be significantly eased.

The first suggestion about "martian oases" (or "swamps") came from Robert L. Huguenin and Stephen M. Clifford of the University of Massachusetts in the late 1970s. Unexplained sources of water vapor were spotted by the Viking Orbiters in two tropical regions, Solis Lacus ("Lake of the Sun," near 25° S, 85° W) and Noachis-Hellespontus (near 30° S, 315° W). The scientists postulated the existence of a near-surface source of seasonally liquid water. Additional supporting data included photographs of fog, detection of thermal readings typical of wet dirt, and some earth-based radar returns that looked like those caused by (and only by) liquid water. Also, major dust storms seemed to spring from that area repeatedly.

Further analysis of ten-year-old radar tapes lent additional support. In 1980, two New England scientists (Peter Mouginis-Mark of Brown University and Stanley Zisk of the Haystack Observatory) released the results of their work on radar data originally collected in 1971 and 1973 by NASA's Goldstone antennas in California. The existence of water at Solis Lacus "stuck out like a sore thumb," they told newsmen. The radar echoes showed a seasonal variation, indicating liquid water in martian summer and ice in winter; such changing signals "can only be explained by the presence of water," Mouginis-Mark concluded. He added, "The mere fact that there is liquid water near the surface and the temperatures go above forty degrees during the summertime must make this a much stronger bet for life forms if they existed on Mars."

One puzzle was that any water leaving the Solis Lacus swamps (and not all geologists agreed with the original Huegenin and Clifford hypothesis) would not return. Instead, it would eventually freeze out of the atmosphere onto the polar caps. The Solis Lacus water should have all been gone many millions of years ago, unless it was being constantly replenished from somewhere. In 1979, Huguenin and Clifford proposed just such a mechanism: leakage from the ice caps themselves.

In this fascinating hypothesis of a "global subpermafrost groundwater flow system," the accumulation of ice caps would cause water underneath the caps to melt from the pressure and be forced into subterranean channels. These are not likely to be open caves but rather porous layers such as fractured rock and gravel. Such rubble does exist on Mars to a great depth because the surface regolith has been blasted by meteor impacts for billions of years, and bedrock is a long way down. This hypothetical underground ocean is held down under a roof of permafrost except in areas of the planet where that roof would melt or break (from meteor impact, for example). This would explain the massive outflow channels which seem to originate at single points, and would explain why the warmest surface areas (where the roof melts entirely in summer) would be *wet*: the region called Solis Lacus. Poetically true to its name, the sun *does* make a lake there, or at least a swamp!

"Subpermafrost groundwater systems on Earth, such as those in Siberia and in the Dry Valleys of Antarctica, can be found at depths of several kilometers and are often characterized by their high salinity," Clifford has noted. Since salinity affects the freezing point of water, the degree of salinity together with the internal heat of the planet combine to determine the possible depth of permafrost. Fresh water would be frozen to a depth of at least a mile, even in the martian tropics. However, the scientists hastened to add, "If we assume the groundwater is saline, . . . [tropical] ground ice will occur only in the seasonally active temperature layer of the top [twenty feet] of soil."

The swamps, then, would merely be small outlets of a planet-wide water system in which the porous regolith, below a mile or so of permafrost, is soaked with saline liquid water to a depth of several miles more. A drill through the permafrost roof would unleash an artesian well that could spew out liquid water without any pumping requirement at all . . . *if*, that is, if such a water system really exists.

At the other temperature extreme from the tropical swamps are areas cold enough for surface ice to form. There are other areas

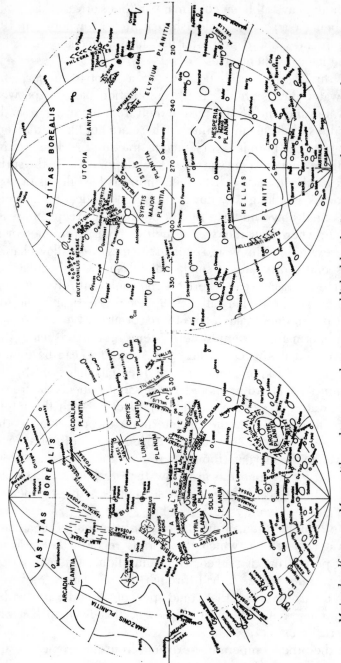

Major landforms on Mars with modern nomenclature as established by the International Astronomical Union.

besides the polar caps, and the most intriguing possibility is the upland region called Tharsis.

On the northwest (that is, lee side) flanks of all the major volcanoes there, but particularly northwest of Arsia Mons, lie weird ridges looking much like plowed fields. The ridges run in parallel and even cross craters. Some observers speculate that they are caused by landslides. But Dr. Carl Seyfert of the University of Buffalo thinks they are something much more interesting. "They could be recessional morrains—dirt left behind by glaciers," he suggested in 1981.

Photographs of the area often show white clouds, which could be light snowstorms or ice crystals blowing over the mountain. These small regions *are* cold enough for surface ice to precipitate and remain year-round. In fact, the flank of Arsia Mons, by virtue of its altitude, is one of the coldest regions on the planet.

Seyfert's theory is that the debris ridges show repeated glacial activity and that the higher altitude ridges, which border on a strange "humpy" topography, are likely to have ice cores. "My gut feeling," he continued to theorize, "is that Mars is just oozing with water" and at Arsia Mons it's just lying around near the surface, as a traditional but dirt-covered mountain glacier.

Since the western slope of Arsia Mons had already been tagged by the Greeley team as *the* most interesting young volcanic target on the whole planet (and there is evidence of on-going activity on the south flank), this combination of *science* and *water* could make it an irresistible target for the first landing which intends to home-stead Mars, use local resources, and stay for a year or two or more.

Using the Apollo analogy again, recall that the second manned landing came down within walking distance of a robot Surveyor probe. Part of the purpose of the flight was to retrieve equipment from the derelict vehicle to determine the effects of long-term exposure to lunar surface conditions; as a serendipitous discovery, still-viable spores were found inside the camera casing when it was brought back to Earth; they had survived on the moon for more than two years.

A similar rationale may bring men back to the Viking lander sites, both to examine nearby areas of scientific interest and to retrieve pieces of the probes to study the long-term (three decades

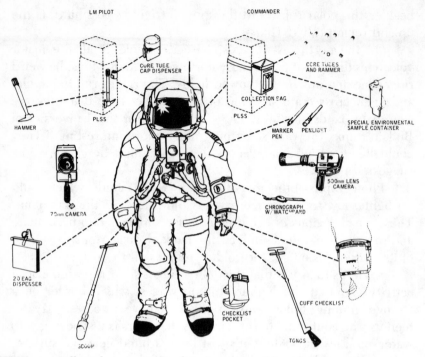

Apollo-style moonsuits and equipment would work well on Mars.

or so) weathering processes on material which might be used for permanent structures on Mars.

One more task would belong to the astronauts on this sortie, a task defined in 1979 by NASA officials who knew only that men *someday* would walk on Mars, but who couldn't hazard a guess at *when*. At a memorial ceremony for a leading Mars scientist who had died while mountain climbing in the Himalayas, Dr. Robert Frosch officially renamed the Viking-1 landing site as the "Thomas Mutch Memorial Station." Frosch unveiled a plaque which he hoped astronauts would someday take to Mars and fix to the Viking probe.

Such variety of possible targets and target strategies should be a constant reminder that Mars is a planet, a whole world. In fact, it has more dry surface area than does our own Earth. A suitable closing thought comes from Dr. James R. Arnold of the

University of California, who in 1979 at the Second Mars Colloquium warned (or promised?) that "Mars is an extremely diverse object and we have consistently underestimated that diversity. I feel certain there are many surprises still in store for us."

Nevertheless, at this point we know plenty about Mars, and it has been enough to formulate some rational, productive exploration strategies at known spots on the surface. New probes will bring new data and possibly new mysteries, affecting the choice of location for Mars Base One, but the logic of its selection will be similar to that already described here.

6

Science at Mars

WE MAY NOT go to Mars for science, but science will be well
served by our going to Mars. And this process, more than any short
range political or social or ideological motivation that actually jus-
tifies the budget, will provide the ultimate payoff in our understanding of Mars, of all hard-surface planets in general, and of our
own Earth—the past, present, and possible futures of all potential
planetary habitats for human beings.

Although planetary scientists may cringe at the comparative
expense of a manned expedition, contemplating the far-ranging
array of unmanned vehicles they might prefer to spend an equivalent budget on, that has never been a genuine option. On Apollo,
the real choice was not between $20 billion of manned exploration
or $20 billion of unmanned probes, but between the politically
possible manned program and a low-priority miserly unmanned
program perhaps a few percent of Apollo's size. Such a program
would have been congenitally unable to retain commitment long
enough to survive, as witness the demise of unmanned planetary
exploration under both Democratic and Republican administrations. Manned exploration of Mars may be the only game in town.

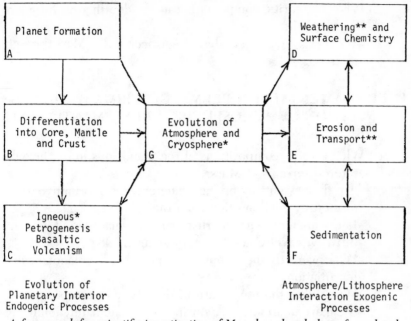

A *framework for scientific investigation of Mars has already been formulated.*

The following questions represent a cross-section of the issues that will be explored as a result of a mission to Mars.

A. FUNDAMENTAL PROBLEMS OF PLANET FORMATION

1. What is the bulk composition of Mars?
   —How does this relate to the composition and physical state of the solar nebula at Mars orbit?
2. What primordial gases are retained in the present atmosphere?
   —What were relative efficiencies for gathering volatiles and nonvolatile materials during planetary accretion?
3. What are the compositions of the natural satellites of Mars?
   —Were Phobos and Deimos formed from the solar nebula in the vicinity of Mars orbit, or were they captured from heliocentric orbits?

4. What meteoritic components can be identified in the martian regolith?

–What classes of objects have been accreted to Mars through its history?

B. FUNDAMENTAL PROBLEMS OF PRIMARY DIFFERENTIATION OF ATMOSPHERE, CRUST, MANTLE AND CORE

1. What are the compositions of rocks and soils in the heavily cratered terrains of Mars?

–Are they essentially primary differentiates of primitive rocks formed by condensation from the solar nebula?

–Is the ancient martian crust primarily made up of igneous intrusives, volcanics, or sedimentary rocks?

–What were the heat sources responsible for planetary differentiation?

–What were the temperature, density, and composition of the martian atmosphere during and immediately following internal differentiation?

2. What are the radiogenic ages of rocks and soils in the heavily cratered terrains of Mars?

–Can we find in martian history a distinct episode of crustal heating corresponding to massive internal differentiation?

3. What, if any, are the remnant magnetic fields of crustal rocks?

–Did Mars have an early global magnetic field or local fields?

–Is the initiation or run down on an internal dynamo indicated?

C. FUNDAMENTAL PROBLEMS OF IGNEOUS PETROGENESIS

1. What are the ranges of composition and age of igneous rocks formed through martian geologic history?

–What were the temperatures, viscosities, and cooling histories of these rocks?

–What causal relationships exist between composition and age, on one hand, and eruption style, tectonic setting, and geographic location on the other?

　　–How many parental magmas are needed to account for all suites of igneous rocks?

　　–How homogeneous, laterally and vertically, is the martian mantle?

　　–How recently were volcanoes active on Mars?

　　–What has been the time history of the flux of objects impacting the surface of Mars?

2. Do martian igneous rocks contain xenoliths?

　　–What is the deep structure of the martian crust?

　　–What is the composition and structure of the martian mantle?

　　–How are radiogenic heat sources partitioned between the crust and mantle?

3. What are the compositions and ages of igneous rocks, if any, which occur on the low-lying plains of the martian northern hemisphere?

　　–When and by what processes, erosional or igneous, was the northern hemispheric lowland formed?

4. What ranges of age and compositions are spanned by volcanic rocks of the Tharsis-Syria Rise?

　　–What events within the martian interior doomed this region?

　　–Are these unique in martian history?

　　–How long has it been volcanically active?

D. FUNDAMENTAL PROBLEMS OF WEATHERING, EROSION AND TRANSPORT PROCESSES

1. What are the physical and chemical characteristics of martian regolith?

　　–What are the roles of water, UV radiation, carbon dioxide, and climate in the alteration of minerals?

　　–Is (are) there typical weathering profile(s) in the martian regolith?

　　–Is there evidence (as fossil soil profiles) of past variations in weathering rate or style?

　　–What contribution do impacting meteorites make to regolith elemental composition, glass content, comminution and transport?

  –What is the distribution of organic materials on Mars and
  what processes destroy them?
  –What is the source of chemical activity in the soils?

2. What, if any, are the distinctive physical and chemical char-
    acteristics of materials in the apparent source and sink re-
    gions of valleys sculpted by flow processes?
    –Has erosion there been mainly by aeolian fluvial or mass
    wasting processes?

3. What are the physical and chemical characteristics of mar-
    tian impact crater ejecta and landslide lobes?
    –Are local bedrock or ground ice characteristics responsible
    for the wide range of morphologies displayed by these
    features?

4. What are the physical characteristics of martian dune de-
    posits?
    –From where and when were rock and mineral grains con-
    centrated into dunes?

E. FUNDAMENTAL PROBLEMS OF SEDIMENTARY
    PROCESSES

1. What are the physical and chemical characteristics of layered
    terrains in the equatorial region of Mars?
    –What weathering, erosion, and transport processes con-
    tributed sediment to these rocks?
    –Are variations in weathering, volcanic activity, or sedi-
    mentation rate, sediment composition or cementation re-
    sponsible for layering?
    –From when in martian history do these rocks date?
    –Is there a useful isotopic or magnetic stratigraphy for com-
    paring sequences of rocks?
    –Do these rocks provide evidence for major changes of cli-
    mate?

2. What are the physical and chemical characteristics of layered
    terrains in the polar regions of Mars?
    –Are these deposits glacial, volcanic, aeolian or other in
    origin?
    –Do variations in volatile content, grain size, cementation,
    composition, or other properties account for the layered
    structure?

—What is the age of these rocks?

—Are they in all probability forming today?

—What distinctly polar weathering, erosion, or sedimentation processes have formed these layered rocks?

—Is there a stratigraphic record of oxygen isotope or magnetic field variations in polar sediments?

F. FUNDAMENTAL PROBLEMS OF MARTIAN BIOLOGY

1. What organisms or organic compounds are present in martian soils?

   —Is Mars now locally populated by any life forms? What environments?

   —Are they continuously or periodically active?

   —What is the fate of organic compounds introduced into the martian atmosphere or soil today?

2. Are there any fossil records of life forms in martian soils or rocks?

   —What, if any, organisms populated Mars in the past?

   —What environments did they populate?

   —From what base did they evolve, local or extra-martian?

G. FUNDAMENTAL PROBLEMS OF THE EVOLUTION OF THE ATMOSPHERE AND CRYOSPHERE

1. What is the volatile content of crustal rocks?

   —To what degree has Mars outgassed?

   —What has been the history of volatile release from the interior?

2. What is the composition of the present martian atmosphere?

   —What distinctive Mars accretional and loss mechanisms have operated?

   —How are outgassed volatile materials distributed among atmosphere, regolith and cryosphere?

3. What is the present dynamical state of the martian atmosphere?

   —What is the mechanism for initiating dust storms?

4. What is the past history of the Mars climate based on the geologic record?

   —Are there indications of changes correlated with orbital and rotational perturbations or the Mars orbit?

–Are there indications of changes correlated with changes
on Earth?
5. What is the nature of the martian cryosphere?
–Are variations in its extent reflected in landforms?

The scientific questions about Mars are fairly well defined. In
1977, the NASA-Houston "Lunar and Planetary Sciences Division"
published *An Outline of Planetary Geoscience*, which contained a
section on Mars, written by Donald Bogard. First, Bogard listed
why Mars was unique. "It is the only solar system object with
surface processes comparable enough to Earth to allow comparison
of similar types of weathering, erosion, transportation, deposition,
and lithification. The solid surface of Mars is the most readily
accessible for scientific study among the inner planets. Its Tharsis
uplift is associated with the largest gravity anomaly known to exist
among the inner planets. Its wind appears to be the dominant
erosional and depositional mechanism at the present time, although
water may have been important in the past. It displays the largest
volcanoes known in the solar system."

To understand these features, Bogard then listed six "interesting and important questions" to be answered by future exploration of the planet and by continuing earth-based observations
and analysis of already gathered space probe data.

1. What is the composition of Mars?
2. Does life exist on Mars?
3. Did liquid once flow on the surface? Is there permafrost?
What are the actual patterns of wind erosion and dune formation?
4. Why does Mars have such large, relatively young volcanoes? How old are they really? Does volcanic activity continue to
this day? Has "continental drift" or other forms of tectonic activity
caused visible surface features?
5. What has been the meteorite impact history? How efficiently have the craters have eroded? How old are the oldest craters? What caused the puzzling regions called "chaotic terrain?"
And where did the satellites of Mars come from?
6. Does Mars have a conducting core?

Another report, by the same team of Houston space scientists,
issued the following year as part of a proposal on unmanned probes,
summarized it this way:

The study of Mars should lead us toward major new insights into the processes by which planets formed; the effects of initial temperatures, pressures, and composition on the subsequent internal evolution of the planet; the time history of internal activity that led to the evolution of its current internal structure, surface features, and the atmosphere; the degree to which impact of meteoritic or asteroidal material may have determined planetary composition and crustal structure; the interaction of solar and galactic radiation with the atmosphere and surface materials of the planet and their role in determining atmospheric evolution; the history and dynamics of the atmosphere and hydrosphere and their relationship with the surface and internal processes of the planet; the nature of the environments in which organic evolution can be sustained; and the resulting biological evolution of the martian surface.

In contrast to what needs to be known, today's state of knowledge is very poor. "The data pertaining to formation, internal structure, petrology, chemistry, differentiation history, regolith formation, and absolute chronology are virtually nonexistent."

Here, according to the report, is where samples from Mars would be of crucial importance. Scientists could analyze chemical and isotopic composition, mineralogy, texture, physical properties, age determination, remanent magnetism, plus the physical and chemical weathering products of rocks and soil. They could calibrate the meteorite flux in the past by the isotopic age measurements on breccias from different cratered regions, and could date lava flows by the same methods. They could identify minerals which must have formed by precipitation from water, or by the evaporation of mineral-bearing water. Lastly, they could search for present, past, or fossilized organisms.

They could do all these things if they had access to the rocks, either brought back to Earth or, in part, examined in place, by robots or astronauts or both on Mars.

One scientific aspect of Mars is both genuine and of great potential popular interest: it has to do with Earth's weather and climate, so its relevance and urgency can quite effectively be communicated.

Space scientist James Cutts addressed this idea in a NASA-sponsored report in 1978, when he tried to grapple with the delicate

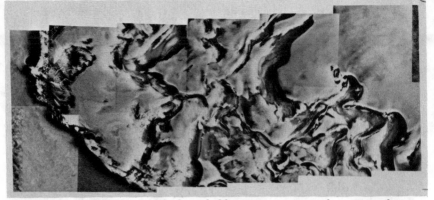

The polar caps have layers which probably contain unique solar system climate records of the past million years, potentially crucial to understanding Earth's Ice Ages and to discovering ways of preventing their recurrence.

problem of public enthusiasm for expensive Mars missions. "An objective is needed which has the flavor of high discovery and relevance. In this era when environmental issues and concerns about planetary survival are paramount, an investigation of 'climatic change in the solar system' might provide the focus that is needed. . . . There are chemical and physical records at accessible landing sites on Mars which can be used to unravel the history of chemical and physical alterations of the martian soil, the age and genesis of certain channels, and the atmospheric processes in the polar regions."

The crucial importance of such scientific findings, Cutts continued, was connected to the fact that Mars "may contain the only independent record in the solar system of climatic effects brought about by extraplanetary factors such as solar variability, which were also experienced by Earth." Mars, unlike the moon, has records of *recent* climate there, perhaps over the past million years or less, and these records could be correlated with similar findings on Earth to search for patterns. If found, such patterns would better allow predictions of future variations.

The Boulder conference endorsed this view, as the summary report stressed. "Geological studies of the martian surface will help us to understand climate change. A study of the climatic record on Mars, as recorded in the surface features and sedimentary rocks,

will help us to better understand the nature of such climatic changes."
Earth, too, is subject to such climatic effects.

The ultimate question of life on Mars also remains to be settled. The remarkable Viking results, while baffling, are now generally considered to represent bizarre soil chemistry but not biological activity. As science writer Leonard David wryly put it at
Boulder, "Viking asked Mars if there was life there, and Mars
replied, 'Please rephrase the question.' "

In 1977, the National Academy of Sciences issued a report by
their Space Science Board, entitled *Post-Viking Biological Investigations of Mars*. The issues involved were summarized this way.

"The discovery and characterization of present or prior life on
Mars would, in the opinion of many, constitute a scientific finding
of unparalleled significance to biology, and it would constitute a
finding of major importance to planetology. . . . Mars is the only
real target for exobiological searches in the solar system. . . . There
are three possibilities for Mars: life exists; life evolved but no longer
exists; life never evolved. The discovery of existing life would be
tremendously exciting. But the other two possibilities would also
represent discoveries of profound importance."

The Academy report then pinpointed another aspect of science
at Mars that could reflect back upon crucial terrestrial concerns.

It is customary to think that life exists only on planets that provide
the proper conditions for its maintenance. But the realization is
growing that life itself may modify a planet's surface and atmosphere to optimize conditions for its existence. Even if it were
demonstrated that life does not exist on Mars, the question would
remain whether Earth and Mars differed sufficiently in their early
histories to permit the origin of life on the former but not the
latter. Or, alternatively, did both planets permit the origin of life
and then diverge dramatically? If so, did the type and extent of
life that evolved play a major role in that divergence?

These questions are of fundamental scientific interest, but
they also may be questions of fundamental importance to all of us
on Earth. We have clearly reached the point where human activities are exerting global effects on the composition of the Earth's
atmosphere and perhaps its temperature. Atmospheric pollutants
may affect the ozone layer and could modify the Earth's albedo.
The burning of fossil fuels has already measurably increased the

carbon dioxide content of the atmosphere, and some scenarios predict serious and even devastating consequences if major fractions of our energy requirements continue to be derived from those sources.

But to understand these processes, one model is not enough. Several different planets on different developmental routes and different evolutionary stages must be studied in detail. That's where Mars comes in, and that's where the issue of life-on-Mars comes down to Earth.

The search for life on Mars may be able to hitchhike on other criteria for site selection, as described in a coming chapter. "Sites that are interesting geologically are also interesting biologically," Jim Cutts told the Boulder conference, listing such tempting targets as volcanic fumaroles, sites for liquid water, and outwash plains from water erosion.

As far as actually searching for life, the presence of an astronaut on Mars is bound to expand scientific capabilities. In 1978, Ben Clark published a paper about manned science on Mars, in which he pointed out that "In spite of the immense complexity and cost involved in miniaturizing the three [Viking] life detection instruments into a simple, compact 'black box', the level of sophistication of these experiments is not beyond what a trained college student could perform with modest equipment on a single lab bench." The following year, astronomer-astronaut Karl Henize wrote how he and his colleagues might detect life:

> Granted that life on Mars may be well concealed just as it is on Earth in hostile environments, how, then, may life on Mars be sensed? As a subtle color shading in dusky crevices? As a pseudo-geometric pattern in a fractured rock? As microscopic hairs in a core sample? The ingenuity of life outstrips man's imagination. Thus life on Mars may indefinitely elude the limited tests and sensors that can be accommodated on unmanned probes. If life *is* suspected, how do we get conclusive evidence? Must we split rocks? Or dig holes of indefinite depth? Or sense sticky spots? Remotely controlled probes may perform some or all of these functions, but one apparatus that could do all would be incredibly large and expensive. . . . On the other hand, human beings can accomplish all these functions and a wide variety of others at costs

which are not impossibly expensive. . . . When the scientific objectives of a mission are complex and not easily preprogrammed, direct human exploration may, indeed, be the most cost-effective approach.

A specific scenario for human exploration of the martian surface was prepared by Ben Clark of Martin-Marietta for the 26th Annual Conference of the American Astronautical Society, held in November 1978 in Houston. The conference theme was "The Future United States Space Program," and Clark's paper was entitled, "The Viking Results—The Case for Man on Mars." Automated equipment was fine, Clark contended (and he had helped design the Viking probes), but real scientific progress would come when scientists themselves arrived at Mars.

> Science progresses ultimately at the hands of a scientist. Sophisticated instruments are merely the tools by which we enhance and extend the fundamental senses we were originally endowed with. Not only must the scientist interpret findings into workable theories, but he must be intimately involved in the observations which form the basis for those findings. Thus, the deduction of geologic field relations among a complexity of units requires a thorough three-dimensional visualization of the terrain, probing of subunits, geochemical analysis, discovery of anomalies and erratics, and geophysical surveys. No field geologist would be content with pictures from a single vantage point and sampling within a very restricted area, if the locale were on Earth. He would, instead, roam the site to quickly establish the general context and then, applying experience and insight, determine which portions of the site deserved extensive study. It is no exaggeration to say that the rate at which a field geologist would cover a given site at a given level of thoroughness would be [a hundred] to [a thousand] times faster than would be possible by remote operation of a rover from Earth.

Clark then went on to characterize an exploration strategy for martian surface features.

> Understanding the mechanism and time formation of the channels will require on-site study, systematic sampling, and possibly deep drilling. . . . Craters are important targets of study since they provide a natural excavation into the crust of Mars, bringing up

materials from possible subsurface layers of differing composition. They also provide natural basins which serve as traps for wind-transported material and for concentration of soluble salts at the topographic low point, assuming liquid water (rainfall?) was once available. A simple but important task would be to study stratification (horizons) within the soil by digging trenches and carefully sampling the sidewalls at regular increments. A very comprehensive program of rock sampling would be undertaken, possibly with the aid of portable geochemical analyzers to seek various representative types. These rocks would then be studied at the 'home' base, using research tools which had been transported to Mars, e.g., thin-section equipment, petrographic microscope, scanning electron microscope (SEM), X-ray energy dispersive element analyzer, X-ray diffractometer, and thermal analyzer. A certain amount of isotopic analysis for age-dating would also be possible. Selected fragments and even whole rocks would be reserved for transport back to Earth, where an even larger range of tests can be performed. As the geological relationships on Mars began to unfold, return visits to sites of particular interest would allow even more discrete study and sampling programs.

Seismic and electrical sounding studies would reveal the presence of subsurface layering and compositional heterogeneities. Setting up and retrieval of the required arrays of sources and detectors would be straightforward for the Mars men, a task anything but simple for automatons. Deep drilling, to 100 meters [330 feet] or more, is quite feasible by astronauts using portable equipment and vastly more important from a scientific standpoint than the one meter drill core currently planned for unmanned spacecraft exploration. . . . During these wide-ranging travels, the scientists would carefully document on film the *significant* features of the terrain. In this capacity, they would provide an immense degree of data compression in the sense of eliminating redundant or inconsequential picture-taking.

Concluded Clark: "The scientific rewards are enormous compared to the limited success that will be achieved if unmanned missions alone are pursued."

The equipment to support such activities can be built. Mars poses its own unique problems, all the same.

After decades of familiarity with spacesuits and spacewalks and moonwalks, the image of spacesuited figures bounding across the martian landscape is not too difficult to envisage. And spacesuits there will be—the martian air is too thin to provide any real-time life support. The astronauts, so dressed, will step out of airlocks and leave footprints in the red dust. They can walk for miles, or ride "Mars jeeps" patterned after the lunar rovers of the early 1970s. Along the way, they will take samples and photographs, set up research equipment, and repair, resupply, and service their operating equipment as needed.

One of the many NASA studies conducted in anticipation of a man-to-Mars program, is a report on a Mars surface spacesuit. The preliminary analysis by engineer Joseph J. Kosmo was published in October 1, 1970, but the only surviving copy of the report that I could find was the spare one which Kosmo had stashed away in the back of his desk drawer when the project died.

The Mars suit was to be a "hard suit" with a two-gas system operating at a pressure higher than that of the pure oxygen suits used in Gemini, Apollo, Skylab, and Shuttle EVAs. The primary rationale for this was to eliminate any need for astronaut pre-breathing to purge cabin nitrogen from his/her bloodstream (too great a pressure drop could otherwise cause aeroembolism, or "the bends"), since the crewmembers would be conducting daily EVAs lasting six to ten hours at a stretch. In 1970, engineers had already collected promising data on the RX suit series (for "rigid experimental"): the suit was based on a sandwich structure with aluminum inner shell and a fiberglass honeycomb outer shell, giving a total suit weight of about sixty-five pounds (the backpack unit would have weighed another 130 pounds).

Differences from the Apollo moonsuits would include removal of micrometeoroid and thermal protective layers. Despite the frigid martian environment, heaters need not be installed. "The metabolic heat output of the crewman can be used for heating instead of rejecting it to the environment," wrote Kosmo. A "buddy system" hookup, allowing two-on-one breathing, probably would also be installed.

According to the study, one key feature of such a suit would be its field maintainability. "Lubricants and spare parts such as O-rings, gas, water, and electrical connectors and possibly spare helmets may be some of the items required. Replacement of various joint systems most susceptible to wear or degradation would be feasible and could be accomplished on the surface by the crewman.

These replacement modular suit parts (e.g., elbow joint assembly, knee joint assembly, gloves, etc.) would be part of the general maintenance kit and be part of the on-board suit spares inventory. . . . A patch kit should be provided in case the fabric in the joints is torn or punctured."

Suit parts could also be swapped out in order to cannibalize the crew's suits. At least two working suits would have to be in near-perfect order to support productive surface excursions, but degraded suits (immobile limbs or high leak rates) would still be usable for short, emergency outside activity.

A decade later, NASA engineers in Houston had practically forgotten that they had ever even thought of an 8-psi "hard suit" (as opposed to the "soft suits" which did not feel like jointed armor). But at the Ames Research Center in California, an advanced hard suit for 8-psi was being developed for Air Force space requirements, in cooperation with the Georgia School of Textile Engineering and some small, local aerospace firms. Suit engineer Vic Vykukal boasted of the suit's flexibility in a report. "The suit uses joints which are true constant volume convolutes," he wrote, explaining why limb movement would not cause the suit's air volume to rise and fall. "It can be placed in any position, from flexed into fetal position to fully extended, and when pressurized will not move at all. . . . There have been some misunderstandings about the ease of use of a glove at 8-psi. The best way to settle this question is to make a direct comparison. Our glove is *better* at 8-psi than the Apollo-type glove is at 4-psi. In addition it is neutrally stable— it can be placed in any position and will have no tendency to move." Such a suit seems best fitted for use on Mars, as NASA's 1970 studies clearly argued.

And such a suit could take surface explorers a long way. Merely walking, they would have an effective radius of about two miles. Riding in one jeep, an astronaut's mobility radius would go up to six miles, a figure set not by the jeep's range, but by the safety restriction of being able to walk back home from a disabled vehicle. With two astronauts, each in his own separate jeep (with an empty passenger seat loaded with experimental equipment), the range would jump to a radius of fifty miles, since a disabled jeep could be abandoned, or, better yet, towed back to the lander for repair. The Jenkins Report highly recommended this option, even for an austere low-budget mission: "The use of two mobility aids increases the potential exploration area by a factor of 100 as compared to that possible with one. . . ."

One of the earliest tasks of the landed astronauts would be to scout their surroundings aboard the jeeps. Since it will take a few weeks for the crewmen to fully regain their muscular strength and endurance in a high-gravity environment (this was the experience of Soviet cosmonauts returning from six month missions in orbit), the more reconnaissance time they spend sitting down, the better. Metabolic studies of the physiological limits of Apollo astronauts indicate that a maximum of two hours driving, followed by at least fifteen minutes at a work stop, is the most efficient schedule. So if the rovers can average ten miles per hour (with top speeds on a straightaway of twice that), the astronauts could range 100 miles or more on a single day's travel.

The Jenkins Report described an even more ambitious sortie, which could be attempted every two weeks or so once the astronauts fully recovered their strength. These would be overnight ventures with the crewmen sleeping in the field (presumably, then, with heaters in their suits). They would have to spend thirty-six hours without opening the spacesuits, but that would allow a total two-day range of about 170 to 180 miles, doubling the radius of action and quadrupling the area accessible.

As it turns out, thirty-six hours inside a sealed spacesuit is *not* particularly difficult. A little-known fact is that the Apollo moonsuit was qualified for supporting an astronaut for a full ninety-six hours straight, in the event of an emergency return from the moon with an airless command module. Several test subjects even endured such a four-day stretch to prove it could be done. Water was obtained from an in-helmet line or from a tube inserted through an access port on the helmet faceplate. Food paste could be squeezed in through the same port. Urine could be drained out; feces would be segregated in a glorified giant diaper. After four days, the crewmen would be alive.

Extension of surface range beyond 170 to 180 miles could be achieved with the use of a portable, inflatable shelter carried on a trailer towed behind one of the jeeps (we are probably talking now about the second or third surface expedition). The sortie crew could drive out for two days, to a radius of 180 to 200 miles, and there set up the shelter. Although conceivably they could make a three-day trip with their suits' capability, safety considerations would rule this out since the astronauts would need to be able to make an emergency two-day return if the shelter deployment were un-successful—and there's your four-day limit.

Various classes of portable shelters have been designed by space engineers, originally for use on the moon but equally appli-cable to Mars. The simplest "tent" could be deployed and broken down repeatedly on a long trip. It would not even have an airlock, but would consist of a flexible bag maybe eight feet long into which the astronauts would zip themselves, then pressurize, and then carefully doff their spacesuits. On the reverse process the following morning, the pressure would be dropped partially to verify that the suits were secure, then dropped the rest of the way to allow exit for the day's activities.

An additional advantage of such a system would be that it could provide emergency salvation for a crewmember whose suit sprung a sudden leak. Even unconscious (the astronaut would cling to life for minutes, even with all air escaped from the suit), the crewman could be stuffed into the bag and saved by a quick pres-surization. Under the worst conditions, the tent plus astronaut could be loaded onto one of the jeeps, and towed by the other one back to base.

Range could be even further extended when more comfortable shelters became available, possibly in caches dropped from orbit at points between 450 and 500 miles from the base camp. Such payloads, including consumable supplies and jeep fuel, would need not weigh more than 3,000 or 4,000 pounds. An expedition might have several of them stashed on the orbiting mother ship, ready for sequential use or for emergency use where needed, including at the base camp itself. The more comfortable shelters included in such kits would be deployable, but would probably not be portable thereafter; astronauts could operate out of such small camps for a week or two.

The traverse to the cache would be undertaken carefully: two days out, set up the tent (if failure occurred, turn back); two more days out, set up the tent (if failure occurred, dash for the cache); if no troubles, arrive at the cache after two more days. Total possible range on the grand traverse: 500 miles.

Ultimately, a system of night-driving (using advanced photo-multiplier goggles to see by starlight and/or latent infrared) with one sleeping crewman's rover slaved (or hooked tandem trolley style) to the driver's, together with three or four days in the suits heading for a pre-positioned and verified cache, could conceivably achieve ranges of well over 1,000 miles. That leg could take at least ten days each way, with several weeks at the remote camp, so the astronauts could expect to be on the trail or in tents for more than a month on each such pilgrimage. But that could open up almost *10 percent* of the planet's whole surface area to access on a single flight!

Merely setting distance records, of course, is hardly worth the trip, unless there are some extremely valuable and unique sites at the end of the trail. Much shorter traverses can satisfy most or all of the scientific and resource questions that need to be answered on early expeditions, but the capability for long-distance travel is there nevertheless.

As on Apollo, the astronauts on Mars will be doing on-site examination of interesting sites, with high-quality maps already prepared years in advance. Some activities, in fact, will be very similar, and core drilling is an example (as Clark's scenario mentioned).

The value of drilling into a planet's surface has long been recognized as part of an exploration-exploitation strategy. It is still being done on Earth, and it was done quite productively on the moon. The Apollo teams carried a thirty-pound rotary percussion drill system (courtesy of Black and Decker) that allowed them to probe nine or ten feet into the lunar surface and extract core samples for subsequent analyses. Heat flow sensors were also installed in deep holes, which helped determine thermal properties of the moon's deep interior and hence its early evolution.

A similar tool is an obvious choice for earliest Mars surface activity, allowing drilling to depths of twenty feet or more with little improvement of the Apollo-era technology. Extracted cores could either be sealed for return to Earth (on-site extraction of the core from its tube, and productive examination of it afterwards, could be difficult); or it could be sampled and discarded (a core tube with evenly spaced holes along its length would allow a parasite drill to push out those small plugs into sample cannisters, while the rest of the core is subsequently dumped onto the ground by just revving up the drill's power head).

Limitation in drilling deeper and more frequently, if an expedition is planning on staying for months (on Mars) instead of days (on the moon), is not necessarily a hardware problem. Power isn't the limiting factor, but lubricant supplies can be. So the availability of locally extracted water or nitrogen is a key factor in opening up essentially unlimited drilling operations on Mars.

Except . . . except for one potential limitation, which also happens to be a factor actually strongly *desired* by many specialists: *permafrost*. It turns out that drills that can zip right through dirt, ice, or rock, grind to a halt when confronted with permafrost (the Apollo drill could do twenty inches a minute in ice or dirt, but averaged only one to two inches per minute in permafrost before overheating). In 1980, the Martin Marietta Company (builders of the Vikings) reported to NASA on the results of a study for a small automated Mars drill system (part of an unmanned sample return mission). "The results of the preliminary simulated Mars permafrost tests were less than successful," the report admitted candidly. But wider research into terrestrial permafrost drilling technology turned up half a dozen candidate systems that would be available if their need was recognized well prior to the actual planetary mission. Surprisingly, although the obvious answer is to melt through the permafrost, the most successful systems (in Antarctica, Greenland, or Alaska's north slope) deliberately kept the drill head *below* the freezing point, to prevent melting and subsequent refreezing.

A few words need be said about the subject of quarantine. Because of the potential for biological discoveries of tremendous importance on Mars, space scientists around the world made an agreement in 1969 that every effort would be made to preserve the biological purity of the martian surface against contamination by spores from Earth—spores which might survive and multiply and later confuse researchers who found them and wondered if they were native or imported. Advanced technologies of spacecraft sterilization were applied to American Mariner and Viking vehicles, and Soviet space engineers are believed to have followed, more or less, the proposed guidelines.

The quarantine period is fifty years. No uncontrolled and unsterilized Earth spacecraft are to land on Mars in that period, which expires in 2019.

Astronauts on the surface of Mars would be a high risk for biological contamination of the planet, unless extremely costly and time-consuming decontamination procedures were undertaken in the airlock during every spacewalk. There seems little chance of avoiding the scientific controversy involved with repudiating that agreement, with the justification that the manned expedition in one year would bring back scientific findings which otherwise would have filled the whole fifty-year-quarantine period, and beyond. The quarantine must be broken.

If Mars is not to be protected from Earth, how about vice versa? Early lunar expeditions underwent an elaborate quarantine process after their return to Earth, a process which was eliminated as unnecessary after two expeditions. Would similar measures be needed for Mars?

The easy answer is no, they would not. The returning astronauts would spend up to a year on the Mars-Earth leg, and any danger should have manifested itself long before they neared their home planet. Once back on Earth, no further trouble need be taken except that required to protect the samples from terrestrial chemical contamination.

To be really certain (or to respond to the remote possibility that some biochemical danger may actually show up on the homeward journey), a post-arrival quarantine option should be available. The best choice seems to be an earth-orbiting facility, and in fact NASA recently published a study of just such a project. It was called "Antaeus," an allusion to a myth about an organism that was kept weak and harmless by being isolated from contact with Earth. Such a facility may be built in the next ten or fifteen years for genetic engineering research; it could serve the Mars expeditions as well, in the event of biological surprises.

Meanwhile, a practical answer, however, must admit that martian material, biologically active or not, could present a significant problem to explorers on the surface of the planet. There is no way that the astronauts can avoid tracking Mars dirt into their spacecraft during their walks outside. That material could be toxic, chemically reactive, or just plain abrasive enough to damage delicate cabin equipment, and for expeditions lasting weeks or months, such potential problems are going to have to be very carefully controlled.

Human footprints on the moon are there for all eternity, or at least for several billion years, whichever comes first. Mars will be different: even under natural conditions, footprints on the surface of Mars would be erased by the subtle but persistent winds in a few tens of millions of years. And the "natural" Mars may only have a few centuries remaining in any case, since descendants of the men and women who leave the first footprints on Mars may be among the future martians who engineer the rainstorms which will wash away those first footprints while leaving an even more permanent and fitting mark of human presence—life itself.

Perhaps some of those first footprints will be preserved in museums and planetary parks, behind protective screens. Few human footsteps anywhere in the solar system will be as significant in the long range.

Ten thousand times a hundred thousand
  dusty years ago,
Where now extends the Plain of Gold,
Did once my river flow;
It stroked the stones and spoke in tongues
  and splashed against my face,
Till ages rolled,
The Sun shone cold,
On this unholy place.

Jonathan Eberhart,
Lament for a Red Planet

7

Living off the Land

THE AVAILABILITY OF economic sources of air and water on
Mars will significantly ease the burden of transportation and mission planning for an interplanetary expedition, even if the astronauts have a perfect closed-loop, recycling, life-support system.
Food supplies grown at Mars would help, too, not only from a
nutritional point of view but also—as Soviet cosmonauts on months-
long orbital missions discovered—from a psychological one.

And the most radical change in man-to-Mars operations would
be if rocket propellant could be obtained there, or manufactured
in some sort of reasonable process. In the limiting case, the availability of a refuelling depot at Mars, together with aerodynamic
braking at both ends of the interplanet route, cuts the earth launch
weight of an equivalent-payload vehicle to a quarter or a fifth of
the "haul-the-propellant" mode. In practice, as such economies

One NASA study for a post-Viking unmanned probe was to use on-site propellant manufacture for the return journey.

become feasible, the total weight may not decrease but the fraction of spacecraft weight devoted to payload may increase by that same factor of four or five.

This is the promise of the martian environment, which the moon cannot provide. As with the expeditionary goals of human explorers of the European Age of Exploration, the crewmen can to a greater or lesser degree live off the land. Certainly not the first time (full there-and-back supplies must out of caution be packed along), but soon thereafter, the material of Mars will help support and expand the range of visiting Earth life forms.

One of the most detailed studies of the use of on-site martian resources was coordinated by Thomas Meyer, a petrochemical engineer associated with the Mars Study Group at the University of

Colorado in Boulder. At the 1981 Boulder conference on Mars, Meyer supported his conclusion that "one possible method of providing some of the life-support materials is to mine the atmosphere of Mars. . . . We have outlined a conceptual design for extracting, processing, and storing consumables from the martian atmosphere to provide water, breathable air, fuels, energy storage, fertilizer, and other chemicals."

Water was the basis for much of Meyer's proposals. But rather than seek local reserves of ice or permafrost, Meyer planned to extract the material directly from the atmosphere. Using data from Viking, Meyer had concluded that a pound of water was contained in a cube of martian air 100 feet on a side. Compressing it by a factor of two would cause half the water to condense out, since despite the thinness of the air and the scarcity of water, the martian atmospheric humidity is nearly 100 percent. Calculations of the electrical cost of this process put it at the equivalent of "about the price of bottled water" on Earth, reported Meyer.

Meyer foresaw no mechanical problem. "A turbine compressor, designed to operate at the low martian pressures, could function on a [day-night] compression cycle with condensation and collection occurring during the night. The solar insolation on Mars, some half kilowatt per square meter, could readily provide the necessary power." However, actual engineering concepts of the weight of the equipment and the mass production rates were not described.

Further compression of the air would produce more water, but at 80 PSI a major step occurs when carbon dioxide liquefies. After separation of the carbon dioxide there remains a mixture of nitrogen/argon (at a 2:1 ratio), with traces of oxygen and carbon monoxide. This could become a buffer gas for breathable air inside the crew habitats.

Oxygen could be obtained by stripping the molecules off of the carbon dioxide via the Sabatier-Senderens process, which uses free hydrogen to make water that can later be electrolyzed into oxygen and hydrogen again. The hydrogen is fed back into the process, but Meyer has worked that out too: "Hydrogen separation can be accomplished chemically or under pressure using iron or palladium diffusion cells," more weight to bring along *if* they are to produce an even greater weight of air supplies.

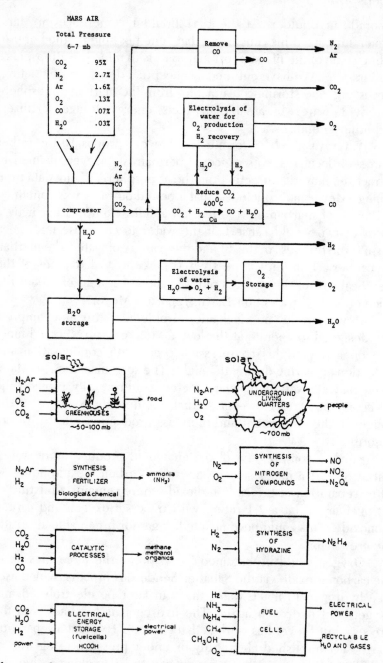

The atmosphere itself can be "mixed" for chemicals which in turn can be processed into highly valuable materials.

The martian atmosphere became a veritable chemist's delight in Meyer's formulation:

> Using nitrogen and hydrogen, ammonia can be synthesized in the Haber-Bosch catalytic process. Ammonia is useful as a nutrient for plants and can be used to produce electricity in fuel cells. With the Raschig process, ammonia can be converted to hydrazine, which works well in fuel cells and is also a constituent of rocket fuel. Ammonia reacted with oxygen can produce NO, which may be oxidized to NO_2. Alternately, NO can be synthesized directly from N_2 and O_2 using an electric arc or a solar furnace at 2100° C. NO_2 can be easily polymerized to nitrogen tetroxide, the other constituent of rocket fuel. . . . It is also an intermediate step for producing nitric acid by the Ostwald process.

Also available and useful would be formic acid, methane, methanol, and other organic compounds. Somewhere in all the plumbing, I'd be surprised if the first martian chemist didn't squirrel away the makings of a moonshine still, too!

Just a note on these processes: of course they are not *free*, since energy must be used to produce the products and a lot of that energy is bound to be lost through inefficiencies. But the value of the fuels is that they provide a storage-medium through which the slowly accumulated energy can be released quickly and controllably, such as in a drag race over the dunes, or in the ten-minute rocket launch which burns fuels it took a year to produce, and is worth it. That still makes excellent logistical sense.

Meyer concludes: "By combining existing technologies we can derive most of the materials needed for life support and surface operations by mining the atmosphere of Mars." At the Boulder conference, other scientists who had suspected something along those lines was possible, but not on the scale envisaged by Meyer, scratched their heads and looked for the catch—and couldn't find any. Their conclusion: Meyer's martian chem lab was entirely feasible.

Now, water can be found elsewhere on Mars besides just in the atmosphere, of course. Leaving aside momentarily the question of ice, there is water locked into the minerals. In a paper published prior to the Boulder conference, Mars Study Group leader Penelope Boston suggested that "an alternate method is the extraction from the regolith of the water of hydration by a simple solar cooker and condenser system. Viking results show that we can obtain one kilogram of water per 100 kilograms of soil heated. . . . These methods of water extraction can be operated continuously, the excess being stored in simple plastic storage bags and exposed to the ambient temperature where it will freeze and remain a ready reservoir in times of need." In other words, an ice-cube maker for Mars!

The next step from physical and chemical processing of martian raw materials is biological processing. An overall strategy for such activity has been outlined by Penelope Boston in a paper summarizing work of the Mars Study Project through 1979. She and her colleagues proposed a careful three-stage development scheme.

First would come the simple algae and bacteria colonies, "which are easily assembled and set in immediate operation to supply certain functions such as waste recycling, gas exchange, possibly food, and eventually fertilizer."

Boston went on. "Well-meaning, but we feel, ill-conceived efforts to provide food in the form of algae, bacteria, and the like are likely to be unsuccessful even if they are shown to be nutritionally acceptable. Lack of palatability, especially on a long-term basis, is probably a distinct psychological debit, and highly undesirable in maintaining a group of people coping with the constant challenge of the exploration effort."

Next would come greenhouses. These could be simple, low-pressure inflatable structures in which the astronauts begin the active cultivation of food crops, via hydroponic techniques utilizing the Mars dirt as a substrate. A plastic tent over a crater would provide pressure midway between surface and habitat.

Finally would come the setting up of a "miniaturized yet complete ecosystem with appropriate organisms at all trophic levels capable of long-term stability." Noted Boston's summary report in 1979: "We feel that a totally mature ecosystem is not profitable enough in terms of net productivity to provide sufficient cullable biomass for the sustenance of relatively high densities of human inhabitants," without at least gross inefficiencies in manpower utilization. "The carrying capacity of the ecosystem can be manipulated within certain limits to provide a compromise situation between high productivity and high stability. Wise use of technological assistance, and perhaps even a shift of emphasis of human consumption patterns . . . may be of some benefit."

The martian soil, it should be pointed out, may not be altogether hospitable for plant growth. Dr. Benton Clark wrote as long ago as 1978 that the Viking soil chemistry results implied that "some pre-processing of the extremely salty martian soil may be necessary to eliminate toxic components." These contaminants, however, need not be all bad. Clark went on to write that once extracted (perhaps just by washing), they "could be used to produce sulfuric acid, halides, cement, plaster-of-paris, glass, metals, superoxides, and any number of important items. The indigenous martian chemistry set could be used to manufacture organics, using carbon from atmospheric CO_2, and from them plastics, paper, elastomers, etc. . . ."—even such disparate items as glue and explosives.

With air, water, and energy available, the next step for long-term Mars visitors is the construction of greenhouses, as mentioned earlier. Two engineering options are open: surface structures using sunlight but entailing the problem of walls and skylights; and underground structures entailing the problem of providing artificial illumination.

For a greenhouse on the surface, a key design feature is the desired internal air pressure. To achieve a copy of Earth's atmo-

Plans for martian greenhouses are based on growing most of the astronauts' food on location.

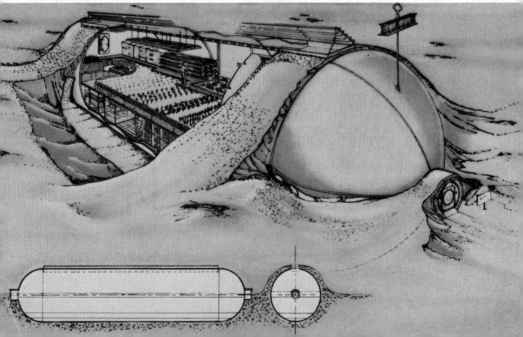

sphere, so people could work inside unprotected, would require excessively strong walls, but earth-side plants do not use most of the natural constituents of Earth's own atmosphere anyway. Lower pressures are often quite tolerable by some food crops. This significantly eases the structural problem, but forces the gardeners into some sort of protective gear to tend their crops.

Estimates were made at Boulder on the size of a small self-sustaining garden for Mars. To support each person, up to half a football field of area could be required; a crew of ten would need a roofed crater 400 feet in diameter. These figures were "surprisingly large" in the words of Richard Johnson, an engineer at the NASA Ames Research Center in California who served as chairman of one of the Boulder workshops, but the full implication of intensive agriculture with high-yield, high-protein crops was not examined.

One compromise design developed by Johnson's workshop involved a greenhouse with air at about one-third of Earth's normal sea-level pressure (1,000 millibars, or mbars). It consisted of a mix of 10 to 20 mbars of carbon dioxide, 300 mbars of a mixture of nitrogen and argon, and about 20 mbars of oxygen. This compares to 6 to 8 mbars of carbon dioxide outside on Mars, or 800 mbars of nitrogen, plus 200 mbars of oxygen, plus traces of carbon dioxide on Earth.

Gardeners in such a structure could work in simple oxygen masks. The solution of the structural problems, such as possibly building the walls and skylights out of local material, were not well developed.

A lower-pressure greenhouse had been designed a few years earlier by members of the Mars Study Group at the University of Colorado. The walls would be a simple inflatable plastic tent, sufficient to hold in thirty to forty mbars of carbon dioxide, five times outside pressure. Several Earth plants were found which could survive and even thrive in that environment (which the Colorado Mars enthusiasts reproduced in "Mars Jars"). The most outstanding one was the radish. "We spent months collecting radish recipes," noted researcher Chris McKay wryly. Some Mars-minded agronomist is going to have to do for the radish what George Washington Carver did for the peanut—produce a hundred recipes for useful foodstuffs and materials.

One of the more valuable, but frequently overlooked sources of martian resources available for immediate utilization is the expedition's own spaceship (or at least that portion used only for the landing). Many structural units could be cannibalized for construction of shelters or exploratory equipment. The parachute, if recovered, could serve as the thermal shroud or even a low-pressure dome over a garden. Descent propellant tanks will probably not be empty, and their chemicals could be used as a bonus power source. These same tanks, once emptied, could then be used as temporary storage facilities for locally produced water, which later would be frozen in big plastic bags piled in the shade. The jettisoned heat shield would hit the ground several miles from the touchdown point of the lander. If found and dragged back to the base camp, it could serve as insulation or structural roof reenforcement for a shelter or garage. The descent engines are likely to have powerful turbine pumps which could be designed for switchover to compressor service after landing. There will be lots of pipe, tow cable and electrical wire, instrumentation, valves, heaters, batteries, electric motors, and other typical space surplus junk, all extremely valuable for use by the Mars astronauts.

There's one more potentially happy coincidence awaiting Mars explorers who intend to live off the land. Water can be broken down into hydrogen and oxygen, excellent propellants for rocket engines; carbonaceous material might be converted into hydrocarbon fuel (perhaps with the aid of bio-engineered bacteria); but one more potential rocket fuel is literally just wafting across the planet's surface: argon.

As discussed earlier, one of the more efficient propellents for SEPS, the solar electric engine (to be used on the interplanetary legs of the journey only) is argon gas. Other designs use mercury or similar conductive materials. One major inducement to developing the SEPS might turn out to be that argon is going to be a by-product of air and water extraction processes at the surface base. *If* an economical way can be developed to get it into orbit (a big *if*), another breakthrough can be expected.

"Argon was the thread—it pulled it all together from our viewpoint," declared Stan Kent at the summary session of the Boulder conference. Describing the results of his workshop on mission strategy, Kent made an even grander prediction: "With Mars as a fuel depot, we are perfectly staged to range the whole solar system [almost out to Jupiter], where solar energy fall-off cuts electrical output," at least for solar cells. Beyond that point, nuclear power plants could take spaceships refuelled at Mars out to and beyond the Galilean moons of Jupiter, where new fuel depots are no doubt still waiting to be discovered or recognized. The argon on Mars could be that significant.

There's an even more exciting option: using *oxygen* as the electric propulsion fuel instead of argon. "Oxygen will work fine in plasma engines," noted Houston space engineer Keith Holden, "and it should work in ion engines too, if scaled-up engines are built." Most conveniently, oxygen can be found throughout the solar system, not just (as in the case of argon) in the atmospheres of Mars and Earth. Oxygen can be extracted from lunar soils, from asteroids, and from the material of Phobos and Deimos—in fact, it is a *waste product* in most "space mining" operations. Such utility is counterbalanced by a potentially serious drawback: "Engine corrosion could be a serious problem," Holden warned. "But if this can be overcome, then the martian moonlets can provide refuelling for electrical engines of interplanetary voyagers" (as they already can provide liquid hydrogen and liquid oxygen for use in chemical engines, too).

What kind of image have we conjured up in this chapter? Mars is very, very different from the moon in natural resources. It is rich in easily accessible volatile materials vital for life support and surface operations. It can provide, on its surface, material which would otherwise have formed (together with the needed weight of propellant) a significant portion of the mass of the departing Mars expeditionary craft back in earth parking orbit. And it is a potentially hospitable site for long-term, if not permanent, inhabitation, but that's still another matter.

Our travellers crossed a space of about a hundred million leagues and reached the planet Mars. They saw two moons which wait on this planet, and which have escaped the gaze of astronomers. . . . How difficult it would be for Mars, which is so far from the sun, to get on with less than two moons!

Voltaire, 1750, in
Micromegas

8

The Phobos-Deimos Detour

ONLY ON HIS third voyage to America did Columbus actually reach the mainland. Only on the third manned flight to the moon did Apollo astronauts actually attempt the landing. Reasoning both by analogy and by unique martian logic, some analysts have recommended that the first manned expeditions to Mars not land on the planet itself, but rather go into orbit around it and visit its two small moons Phobos and Deimos.

One leading advocate of an early manned visit to the martian moonlets is astronomer Fred Singer. In honor of the target objects and of the advanced scientific nature of the proposed project, Singer has nicknamed his proposal the "PhD Mission."

"There are some scientists who would regard Phobos and Deimos as even more interesting objects than Mars itself," wrote Singer in the initial report of a small study project sponsored by NASA in 1977–78. "Their origin is a real mystery. No theory has yet explained their existence. We cannot tell whether they were cap-

149

tured or whether they were formed in place at or near their present positions." Because of their small sizes, "they may be remnants of original accretions of the solar system," that is, the long-sought but still undiscovered "Rosetta Stone" of early planetary formation and evolution.

Singer's advocacy of *manned* planetary exploration came as a significant change of mind. "I was always a strong proponent of unmanned exploration throughout my career," he told the Boulder conference, "but in the past fifteen years I have transitioned. I now believe that man can do it better, cheaper, and sooner than automata." Some of the mind-changing rationale, according to Singer, was practical politics. "Such a major program cannot come into being without public support," he wrote in 1978, ". . . but scientific results, no matter how valuable to scientists, are difficult to convey to the public. It is almost axiomatic that the presence of men and women in the exploration process can create public excitement and involvement: in other words, a major mission must be a manned space mission."

The resulting PhD plan calls for a 600,000-pound vehicle in low earth orbit, which would use a combination of chemical (liquid hydrogen fuel) and electric (SEPS, with argon as propellant) engines. A crew of eight specialists (six planetary scientists and two medical specialists) would be aboard. They fly to Mars orbit and "land" ("dock" would be a better term, considering the almost negligible gravity) on Deimos, the outermost moonlet and thus the more accessible one in terms of velocity requirements.

Once established on Deimos, the astronauts dispatch a series of robot rovers and simple penetrators to the martian surface. Since Deimos is close to a "synchronous" altitude above Mars, it remains within line-of-sight of surface points for up to forty hours at a stretch. This allows pairs of remote-controller astronaut-drivers to control rovers and research stations continuously over these periods (a small communications relay satellite could support simple status reporting during other periods).

This is a major benefit of having humans near Mars: their direct control of a fleet of rovers would be far more productive than either remote control from Earth (with round-trip commands taking hours to form, process, and execute) or by necessarily simple-minded guidance computers on the rovers.

A second major benefit also immediately follows: the Deimos explorers would have a dozen or more landers at their disposal and would not land them all at once. Instead, they would hold some back and send them down to areas of particular interest discovered by the first wave.

At least some of the rovers would be equipped with sample return rockets which would fly up into orbit (perhaps all the way to Deimos, perhaps just high enough to achieve a stable orbit and wait for pickup), later to be retrieved for study along with samples from Deimos itself. In addition, Singer's plan envisioned a three-week, two-man side trip inwards to Phobos for more sampling (that flight could also be assigned the task of picking up the sample return cannisters launched automatically from the martian surface).

After two to four months on Deimos, the mission would depart for Earth, leaving their laboratory-habitat for possible future occupancy. The whole expedition would last about two years.

Approximate costing estimates were done with the help of NASA analysts. Singer came up with a price of well under 10 billion 1978 dollars, or less than one-fifth of what Apollo cost in the 1960s. He also told the Boulder conference that such a mission would be feasible early in the 1990s if funded soon.

During his presentation, Singer stressed that the manned mission would return more science per dollar than an equivalent expenditure on unmanned missions, while having the advantage of being able to attract far more public enthusiasm and support. "By comparison with the manned landings on the surface, the manned landing on the satellites is easier, far less costly, safer, and can be done much sooner. It is also scientifically more productive because it allows the efficient control of a large number of rover vehicles at different locations on the martian surface. It therefore allows sample recovery from different locations, which can be selected one after the other in sequence, depending on what has been discovered from earlier samples. Also, it allows the use of far better instruments for examination [at Deimos Base] and therefore no deterioration of the sample."

Two additional mission goals for PhD came out of the Boulder conference: landing site reconnaissance and on-site resource utilization. A powerful telescope emplaced on Deimos would allow visual and photographic surveys of portions of the martian surface down to one-meter resolution, just good enough for certification of future manned landing sites. And the carbonaceous nature of Deimos (and Phobos) makes it an ideal candidate for raw material for conversion into water, air, and even hydrocarbon fuel for rockets and surface rovers. (Singer, a long-time specialist in the moonlets, allowed that he was "very willing to sacrifice a little bit of Deimos for resource utilization studies, although I do feel a little possessive of [it].")

A science writer at the Boulder conference suggested, only half in jest, that Singer rename his project. "PhD sounds too elitist," Dick Hoagland pointed out. "I suggest 'Project Swift,' since it would be the swiftest potential man-to-Mars expedition—and that would also commemorate Jonathan Swift, who predicted the existence of the two moonlets in *Gulliver's Travels* a century and a half before they were discovered." Additionally, of course, some acronym-brained space type might be able to figure out a meaningful phrase that S-W-I-F-T could be made to stand for, but nobody at Boulder bothered to think of one.

Initially, the Boulder conferees seemed to consider Singer's renegade PhD mission a detour, a dilution of man-to-Mars resources. But over the course of the colloquium, at the give-and-take roundtable discussions and in small groups over meals, the consensus shifted perceptibly: a Deimos manned mission might indeed be a valuable intermediate step on the way to the surface of Mars.

As a precursor expedition, it certainly reduced the size of the bite required to accomplish the actual landing. Many concerns that otherwise would have bedeviled the landing expedition, such as medical effects, or the uncertainty of on-site resource utilization, would have been already answered by the time of the landing mission. The astronauts' mission module and the propulsion modules would have been verified under actual flight conditions, although possibly without the highly sophisticated, fully closed-loop life-support systems still under development at the time the PhD would have to be launched. Highly desirable unmanned precursor missions (such as surface sample return and high-resolution mapping) would have been completed by the manned PhD mission on a scale and precision unimaginable for automated systems. And sophisticated facilities, including an emergency shelter on Deimos and a relay satellite or two, would be already emplaced and waiting when the landing expedition arrived.

It would not be out of the question for a follow-up Deimos expedition to set up a semi-permanent occupancy. Emplaced deep into the interior of the moonlet, such a base would be secure from the severest solar flares and cosmic rays. Chemical and physical processing of Deimos dirt could produce air and water, and greenhouses on the surface could grow significant food supplies (Deimos has about thirty hours in one whole day).

Flight scheduling was a question not seriously approached at Boulder. The PhD mission should probably precede the landing by at least two Mars launch windows, or more than four years—probably quite a bit more. An expedition coming back two years after launch would allow for only a few months overlap before the follow-up expedition departed, so the second expedition would have to be organized primarily on data radioed back from the first expedition, which argues for sophisticated analysis apparatus on the first expedition, including optical and electron microscopes, a mass spectrometer and a wide variety of other radioactive sensing devices and age-dating equipment. Only by the third expedition (each add-on mission costing perhaps one-quarter as much as the first mission) could planners expect to have digested the sample analysis data from the first visit (verified by work on the second mission) sufficiently to prepare equipment designed for operational and reliable on-site resource extraction. So if the third expedition (perhaps up to a dozen people) is tasked with the establishment of a semi-permanent Deimos base, then the manned landing might not be practical until the next window, six and a half years after the departure of the first man-to-Mars expedition.

As targets for exploration in their own right, the moonlets Deimos and Phobos are certainly intriguing pieces of planetary real estate. The most sensational recent speculation about them involved a mysterious "secular acceleration," or unexpectedly rapid decay, of the orbit of the inner moon, Phobos. This led to a hypothesis by Yosef Shklovskiy that the object's unusually high drag was due to the fact that it was hollow, and hence obviously artificial. But that theory, which received very wide popular circulation, was based on what turned out to be incorrect notions about the mass and stiffness of both moonlets. With spacecraft data from the 1970s, astronomers found that tidal drag interaction with the equatorial bulge of Mars could adequately account for the orbital deviations. They did recompute the decay rate and found that Phobos will hit Mars in the relatively brief span of thirty million years.

The objects are highly irregular in shape ("like a diseased potato" in the words of scientists who first saw them in photos in 1971): Phobos measures about seventeen by fourteen by twelve miles, with a volume of 1200 cubic miles and a surface area of 400 square miles; Deimos is smaller, measuring nine by eight by seven miles, with a volume of 240 cubic miles and a surface area of about 150 square miles. Surface gravity is one-thousandth of Earth surface norm, and escape velocity is about twenty feet per second—within the range of a hand-thrown object.

Their composition can be inferred from their color (very dark gray, with an albedo, or reflectivity, of less than 6 percent) and density (about two grams per cubic centimeter, computed from mass values deduced during close fly-by passes of Viking Orbiter spacecraft in 1977). The resulting candidate is material similar to that in two types of carbonaceous chondrite meteorites, which contain a large proportion of water (up to 20 percent by weight), some carbon (but very little nitrogen), along with other elements and compounds potentially useful to space explorers.

The moonlets' surface texture could provide a protective habitat for visitors, since specialists have estimated that both objects are covered with loose soil (or "regolith") to a considerable depth—several hundred feet on Phobos and up to 100 feet on Deimos. Pictures of the moons show that Deimos appears smoother, but the crater counts on both moons are similar, with the apparent difference being that craters on Deimos are more filled with the surface debris. That makes Deimos' craters ideal places for visitors to burrow into for protection and for mining.

One unexpected feature on Phobos, revealed by Viking probes, is its surface system of grooves that rings the moonlet and which is evidently connected with the formation of the three-mile-wide crater called Stickney (named after the wife of the discoverer of the moons, who encouraged her husband when he was ready to give up the hunt). Deimos does not have grooves, probably because there is no crater large enough to nearly have shattered the moonlet, as seems to have happened on Phobos.

Crater chains on Phobos may be steam vents from the interior of the moonlet, and as such, may contain relatively pure deposits of water ice.

Along many of the Phobos grooves are raised-lip craters strongly suggestive of outgassing, or the escape of vapors from within the moonlet. It has been seriously suggested that the impact heat of the Stickney parent object was sufficiently high to vaporize water trapped within the interior rocks of Phobos, the water steam then percolating to the surface along the fractures. If this is so, trapped water ice in a reasonably pure form could well exist on Phobos inside these crater vents, and the water could be readily found, mined, and utilized by thirsty visitors from Earth. Water, or other material, could be a source of rocket propellant as well.

One implication of a fuel production facility on Deimos (or possibly Phobos) is that a nearly totally reusable space-to-surface-and-back transportation system could be set up for Mars. Such a shuttle system could support large colonization activities on the surface, or the export of surface resources (such as argon fuel for electric propulsion engines) to space. The economy is obvious, and the presence of the moonlets is the key.

The system could consist of a fifty-ton propulsion module, including forty tons of propellant and a small crew cabin, attached to a ring of cargo modules (carrying ten to twenty tons of equipment), a large expendable heat shield covering the base of the assembly (with a hatch for the engines to fire through), and a fifty ton auxiliary propellant tank module docked to the front end of the assembly. The propellants—liquid oxygen plus either liquid hydrogen or some hydrocarbon fuel—all have been refined from raw materials on the moons.

Half of the auxiliary tank's propellant is used to generate the 2,200 fps needed to enter a "trans-Deimos" ellipse orbit that dips through the outer reaches of the martian atmosphere, where air braking subtracts another 3,600 fps and allows the ship to maneuver into a low circular orbit. The auxiliary tank is then separated, and the lander makes its atmospheric entry and propulsive-braking touchdown. The cargo modules are detached, up-cargo is loaded, and the propulsion module lifts off (possibly within a day of landing), using up most of its own propellant to achieve a low orbit. Once there, it performs a rendezvous and docking with the half-full auxiliary tank, whose propellants are sufficient for the two burns needed to reach Deimos and the refuelling station there. After a few days or weeks, the ship could go down to Mars again and again.

The bottom line is that mission planners should never forget that Mars is a system of three objects, not just the big red planet. Two nearby moonlets are both more accessible to visitors and composed of extremely useful materials; their purely scientific value cannot be ignored, either.

The disappointment of crewmembers who travel hundreds of millions of miles to the martian vicinity without being able to cross the last few tens of thousands of miles to the surface is certainly understandable, but few should pity them. There's plenty of challenge and plenty of excitement on a Phobos-Deimos mission, even without Mars. They are worlds in their own right, and they have a prominent and too-long-ignored role to play in support of the manned expeditionary programs involving Mars itself.

Not one penny for this nutty fantasy.

Senator William Proxmire,
1978

9

Cost

NOW COMES THE part about money. How much is it all going
to cost, and will any single nation be able to afford the price? How
trustworthy are these estimates likely to be?

In the introduction to this book appeared the startling claim
that an initial manned Mars expedition could be carried out for a
price tag small in comparison to the Apollo lunar program of 1961–
1972. Whatever criteria one chooses—adjusted inflated dollars,
percent of gross national product, cost per capita, and either total
or peak year funding—*all* the careful estimates run between a half
to just over a full "one unit Apollo expenditure equivalent," or
"one Apollo."

That sure doesn't jibe with widely published so-called estimates, which ranged up to $100 billion in 1970 currency (four
Apollos), perhaps a quarter of a *trillion* dollars by the time the
project need be formally funded in the later 1980s. White House
science advisor Frank Press promoted such overestimates when he
opined in 1977, "If the Russians want to spend $70 billion to get
to Mars in five years, we say, 'God bless 'em.' "

The origin of a super-high price tag for man-to-Mars can be traced to the mid-1960s when a sequence of scientists—none with direct experience in space engineering—issued what space workers affectionately like to call SWAGs ("Scientific Wild-Assed Guesses") on the expenses, based on strained analogies with what they thought they knew about Apollo. In 1965, Dr. D. F. Hornig, science advisor to President Johnson, told a Senate committee, "If we compare the probable scale and technical difficulties of a manned Mars expedition with Apollo it is hard to conclude that its probable cost could be much less than perhaps five times that of Apollo—that is, of the order of $100 billion." The same year, Abraham Hyatt wrote in *Astronautics and Aeronautics* magazine that "the cost of a manned Mars expedition is estimated at $75.2 billion spread over a fifteen-year period." Dr. Harry Hess, chairman of the Space Science Board of the National Academy of Sciences, was getting close when he told a congressional committee that "it would cost somewhat more to go to Mars than to the moon, but more likely a factor of two than a factor of ten." But a Douglas Aircraft Company study, released in June 1965, put a price tag of $40 to $50 billion on the project.

Wait a minute. There is a hint in that Douglas study, when you look at how it was broken down. It included $15 billion for development of a new launch vehicle, $3 billion to build a space station, $6 billion for supporting precursor unmanned probes, and $3 billion for development of a nuclear rocket stage. The actual specific man-to-Mars space vehicles were estimated to cost just $20 billion, or one Apollo, a number that was beginning to be genuine because it was based on real aerospace engineering studies, not SWAGs.

Among other analysts, who actually did their homework on estimating costs, firmly grounded price lists were produced. In 1966, Krafft Ehricke gave his view. "It is simply not possible to make an unqualified and general estimate of *the* cost of *a* manned planetary mission. From the study of a large number of missions and associated conditions it was found that the cost of the first mission lies anywhere between $5 billion and $25 billion. . . . One important reason for the relatively low cost of developing a manned planetary capability is that . . . many of the preparations are identical with those required to attain a better orbital and lunar capability."

Ehricke had, as usual, put his finger on the precise point which other analysts would later come to recognize as the factor reducing the $100-billion SWAGs by a factor of five or ten. The *new* technology needed for man-to-Mars was nowhere near as extensive as was the leap from standing still to a man-on-the-moon capability, and much of it was being developed anyway, even without Mars in mind.

Based on studies conducted by Dr. Charles S. Sheldon II in the early 1970s at the Science Policy Research Division of the U.S. Library of Congress's Congressional Research Service, a very similar picture emerged. ". . . If one assumes a [space] program will have other reasons to develop a reusable shuttle, a versatile space tug, and a universal space station module (all to serve many Earth orbital economic, military, and scientific purposes) then even the total costs of developing a Mars expedition become far different from the kind of $100 billion figure which has been common to the literature. People tend to overlook how much of the Apollo costs were associated with building a basic U.S. space capability rather than just going to the moon per se. One might think of a Mars expedition of the type discussed as much closer to the order of magnitude of $10 billion rather than the $25 to $35 billion of Apollo or the $100 billion postulated so often for Mars."

Similar figures were generated by other independent studies. The 1971 Jenkins Report at NASA's Houston space center computed costs for both chemical and nuclear "austere" Mars expeditions, in 1970 dollars: the former would cost $6.925 billion for the first mission and $1.430 billion for each additional mission; the latter would cost $8.725 billion for a first mission and then $1.280 billion for subsequent flights. In 1977–78, Fred Singer's plan for a Mars orbit (no development of the critical and expensive landing module) costed out at a one-fifth Apollo; adding the Mars surface crew-carrying lander might double that price.

In the late 1970s, analysts in the British Interplanetary Society, based in London, did their own cost studies for a joint USA/Europe man-to-Mars expedition. Engineer Robert Parkinson reported this result and rationale in a 1981 paper which evolved entirely independently of the Boulder movement. "One of the reasons that I have been at pains to emphasize the extent to which hardware required may be available is to show that the costs for such an expedition are largely a matter of accounting. . . . If we cost only those items specific to the mission the cost is modest—about $4 billion. . . . *Given the right circumstances it is actually cheaper to send men than to try to do the same thing with dozens of robot expeditions.*"

The validity of these estimates, of course, can be accepted with skepticism in a world where "cost overruns" appear on almost every newspaper and television news broadcast. The aerospace industry, to be fair, is hardly unique, however justifiably it has on occasion earned criticism—consider by comparison San Francisco's BART, the Washington DC METRO, the Congress's Rayburn House Office Building, New Orleans' Superdome, or any number of similar budgetary nightmares. However, cost estimates for the Mars project, when not part of competitive bidding, can be based on reliable, analogous experiences, and such estimation techniques were good enough to track the cost of the Apollo program years before its completion on time and within the allocated budget. Aerospace cost forecasters have nothing to be ashamed of in such cases.

But why so cheap, one asks again. There is yet another reason why a Mars spacecraft in the 1990s would be *less* expensive than even the reliable estimates made in the 1960s: the advance of technological capability and experience, and the relaxed need for "uncertainty margins" as the "uncertainty" associated with a Mars mission fades away.

Humboldt Mandel, a space engineer in Houston, put it this way in a paper prepared for the Boulder conference.

> Advances in structural technology will make the structural masses required only a fraction of those estimates in 1960. The weights of the avionics components required to guide and manage so ambitious a flight have been reduced to negligible percentages of their 1960 weights. Even more significantly, the reliabilities required for long duration flights are now attainable within current states of the art in avionics, and at negligible increases in subsystems weights. Other advances in the weights of electrical equipment, motors, relays, and power generating devices have made the planetary vehicles of today's design bear little resemblance to those of the 1960s.

Based on Apollo and Shuttle experience, Mandel had first produced an intuitive break-out of the program costs (1980 dollars): for the in-flight Mission Module, $5 to $8 billion (and possibly much less if it is based on a 1990s-era space station module); for the Mars landing module, $3 to $5 billion; for the propulsive stages, $3 to $5 billion (and possibly less if the stages have been developed for another program); for the preparatory ground operations, $1 or $2 billion; for the flight operations and transportation costs, $5 to $10 billion per mission. And how much does it all add up to? Mandel did the arithmetic. "Well, first, even adding room for cost growth it represents much less than paid for the first manned lunar landing. In terms of contemporary Gross National Product equivalents, it is only a fraction of what Apollo cost."

At Boulder, Mandel published updated precise numbers generated by use of previous space vehicle costing experience. He made the following assumptions: the spacecraft used chemical propulsion, with solar power, weighed 3.9 million pounds in low-earth parking orbit, took five astronauts on a 600-day expedition (with forty days on the surface); all costs were in 1981 dollars, with an eleven-year development cycle, no major new facilities required (except an operational manned space operations center, justified on other grounds), and use of NASA economic parametric models verified during Apollo *and* Shuttle (despite bad press about overruns, Mandel said Shuttle was only 3 percent over a budget modified to include program change decisions from the White House).

The result was a slight variation on his intuitive numbers, but with a great deal more dependability. The Mission Module costs out at $1.8 billion. The landing module requires $2.6 billion. A new earth entry module costs $800 million to develop (a figure similar to one which prompted the Jenkins Report ten years earlier to find an operational way around requiring such a development). The propulsion module costs $1.6 billion. The transportation to low-earth parking orbit costs $6.7 billion. The experiments and crew equipment have a price tag of $3.1 billion. Institutional support, administration, and ground testing ring in at another $3.7 billion. The total of $20 billion 1981 dollars compares with the Apollo unit, $63 billion 1981 dollars, and with the Space Shuttle development cost of $14 billion 1981 dollars.

The costs can be compared in other ways. As a percentage of contemporary gross national product, Apollo cost 2.8 percent, the Shuttle 0.5 percent, the Mars program 0.6 percent; as a per capita expense, Apollo was $325, the Shuttle is $67, and the Mars expedition would be $76; as a percent of national budget, Apollo was 2.2 percent, Shuttle is 0.4 percent, and the Mars flight would be 0.8 percent; in terms of peak annual funding, Apollo hit more than $9 billion, the Shuttle $2 billion, and the Mars project would hit $2 billion during development and $3 billion during the launch year.

"We've convinced ourselves that going to Mars is such a tremendous undertaking," Mandel editorialized at Boulder, "that when you see real numbers you tend not to believe them. The costs of landing men on Mars, by any measure, are only a fraction of those experienced during Apollo."

One logical consequence of these low cost figures is that a single nation *could* afford to go to Mars alone, thus avoiding forcing the mission to be an international project, which could delay it indefinitely until the arrival of some diplomatic utopia. At Boulder, space newsletter editor Leonard David voiced sentiments independently arrived at by other analysts (including this author): "Such a mission could indeed be afforded by an individual country." The United States and the Soviet Union both could afford such a project, as could Japan, the European Space Agency (or even France, or Germany, alone), or a hypothetical OPEC Space Agency, or perhaps other forces. Arguments that the mission *must* wait for total peace on Earth are not based on reality.

The budgetary savings of an internationalized mission might be more apparent than real (this is not to disparage the political and societal value of such a cooperative endeavor, factors which would probably dominate the arguments in favor of international joint expeditions). Even if the development and operational costs could be split evenly (say, $10 billion each for Washington and Moscow), additional manpower and monitoring efforts would have to be expended on the difficult "interface" function, as well as on mutual assurances that each side was performing its fair share of the task in a safe and productive manner. Such extra expenses engendered by the multilateral nature of the project could run into the billions (warning: this is a SWAG), pushing the cost of "half" the mission well upwards of 60 to 75 percent of the cost of the "whole" mission.

To summarize the rationale behind these far lower than commonly expected costs for man-to-Mars, we have to realize where space technology will be in the post-1999 era. Before going to Mars, Earth's spacefaring nations will already have:

1. bought and paid for the Space Shuttle and its follow-on vehicles, plus probably a shuttle-derived heavy cargo vehicle and possibly an efficient, powerful orbit-to-orbit tug;

2. obtained data on human response to weightlessness of durations beyond one year, maybe two, in Earth orbit;

3. accumulated operational experiences with partially closed-loop regenerative life-support systems in space;

4. coped with medical emergencies in space, some possibly crippling or even fatal;

5. had manned vehicles in geosynchronous (twenty-four-hour) orbit (or "GEO"), which region experiences interplanetary conditions of radiation, magnetism, and micrometeorites;

6. developed a fast-return atmospheric entry module ("advanced Apollo" type) to serve GEO missions and Soviet manned lunar missions;

7. inventoried the material and energy resources of the moon and Mars with cheap robot vehicles;

8. flown unmanned spacecraft into and through many planetary atmospheres, including the use of aerobraking;

9. gained experience with significant solar power arrays and large antennas in near-earth orbit;

10. probably landed semi-autonomous (Soviet) rover robots on Mars;

11. probably used both solar electric and light sail propulsion technology for small unmanned probes, accumulating years of in-flight experience;

12. probably learned to forecast solar flares;

13. possibly developed useful anti-radiation drug treatments;

14. possibly developed a significant on-orbit nuclear propulsion and power system (USSR, not U.S.).

At the same time, certain breakthroughs are not required, but if available (again, with their development billed to some other project) would further reduce the cost of man-to-Mars. Examples are "wild-card" exotic propulsion systems, or a breakthrough in regenerative life-support technology.

The greatest impression was made at Boulder when Mandel's talk was concluded, and half a dozen different experts joined in on an impromptu panel discussion of cost credibility. Again and again, the same refrain sounded: "I wanted to believe my low costing results but they seemed so different than expected—that is, so cheap—that I worried something was grossly wrong." But the confident corroboration each contributor received, from discovering that other people had independently come up with comparable numbers, was heady wine for Mars enthusiasts long accustomed to living in the closet, leaving results in desk drawers, and coming to doubt their own engineering competence.

So, who is now willing to pick up the tab?

To Mars!

Fridrikh Tsandertsander
Soviet Space Pioneer, 1924

10

Are the Russians Going?

WIDESPREAD RUMORS HAVE it that the Russians are headed for Mars—*red expeditions to the red planet.* The direction of the current Soviet manned space station developments may be one hint, but Russian space experts have repeatedly issued explicit public statements in which manned Mars missions are matter-of-factly discussed as legitimate, even inevitable, eventualities.

Chief Soviet space doctor Oleg Gazenko told a press conference in late 1980 that "It is difficult to give an exact date for a flight to Mars. But I think the basic prerequisites for such a flight exist now. Whether the flight happens in ten, fifteen, or twenty years, I cannot say. But I believe it will be before the year 2000." Cosmonaut Georgiy Grechko reported that "Soviet specialists would not be surprised if men land on Mars [in the next twenty years]." In late 1979, cosmonauts Ryumin and Beregovoy reckoned that a manned Mars flight would occur within the next ten to fifteen years. In mid-1980, Yuri Zaitsev, a department chief at the Institute of Space Research of the USSR Academy of Sciences, wrote on the subject. "Taking a look at the future, one can express a firm conviction that the launching of the Salyut-6 station has paved the way for manned flights to other planets of the solar system. . . . A considerable increase in the duration of manned space flights produces an economic effect and helps one to realistically assess the feasibility of manned space flights to the closest planets. In the future, and perhaps in the not too distant future, such flights will no doubt be accomplished."

French experts who have specialized in the Soviet space program have reported that a special man-to-Mars task force has been in existence since 1972. At that time, Soviet space designer Vasiliy Pavlovich Mishin, director of the Central Institute for Scientific Research on "Medium Machine Building" (a euphemism for the rocket industry) began to recruit a team of engineers and technicians to develop a module for manned interplanetary flight. Mishin, who also holds the rank of full academician in the USSR Academy of Sciences, reportedly got the additional responsibility of crew training. The 1981 report by Albert duCrocq in *Air et Cosmos* magazine concludes: "The planets are already at our disposal. They have been waiting for cosmonauts who will overfly them. How much longer will the waiting go on? The time may be shorter than you imagine."

Cynical observers may harbor gnawing doubts that the cry of "the Russians are coming" may be mainly inspired by a desire to lure the U.S. government into a new round in the "Space Race," and commit substantial funds toward an American counterpart. Or perhaps some subtle maneuvering is going on in order to build up the U.S. space capabilities, so that a joint U.S.-Soviet Mars expedition becomes feasible in the year 2000.

The cynics could make much of such candid comments as the one from Congressman Bill Nelson, a Florida Democrat, who talked to a local Cape Canaveral newspaper in mid-1979. "If we find out officially that they are going to Mars it could very well help the U.S. space effort. It would refocus the nation's attention on man's achievements in space, and would inject some competitiveness which would make our job in the Congress of funding the space program easier. The acquisition of funds for the U.S. space program could be considerably eased if the Russians formally announced they are going after a space spectacular, like a manned Mars mission." But would the Russians be so cooperative in funding their competition, one might ask?

What kind of answer is possible to the question "Do the Russians plan to send men to Mars?" No doubt they will want to do so at some obscure, indeterminate point in the future, as a logical development of what they see as their role of the leading social system on Earth. And, as in the U.S. during the Apollo heydays, they almost certainly are funding engineering studies of requirements for manned interplanetary missions. But how soon, and how serious? If they are looking at the mid to late 1990s, then it's obvious that no ironclad policy decisions need even be made for many years to come. More pressing prospects involve large space stations in low earth orbit, geosynchronous orbit, and even lunar orbit. If however, the Soviets are looking at extremely rapid development of an early capability for manned interplanetary missions (perhaps only fly-by probes for the first several years) in the late 80s or early 90s, then obviously some hard budgets exist and some hard-fought political battles must already have been won. What possible traces of such developments would be visible to us?

Here's what can be estimated. The Soviets are acting now, in their space mission development, just the same way they could be expected to act if such urgent man-to-Mars plans did in fact exist. Worded another way: from what we see of their manned space program, there are no obvious "missing links" regarding areas of technology which would *have* to be developed before manned Mars expeditions would be possible six to ten years from now. The telltale building blocks are all there or on their way, perhaps justified for other more urgent programs, but there nonetheless, available for diversion to any man-to-Mars mission planned around 1990.

The biggest puzzle is this: Would a long-range Soviet manned space program which does *not* include man-to-Mars on its agenda *still* have use for all of the building blocks which are appearing. Frustratingly, the answer is probably yes, so we cannot make a definitive answer to the main question asked earlier. But let's examine the thrust of Soviet manned space technology in order to map out, if not their intentions, then merely their impending capabilities. These are impressive indeed: a habitable module that has already demonstrated a four-year-plus flight time (Salyut 6); an Earth-landing module that has long ago been flight-tested at interplanetary return velocities (the Soyuz variant code-named "Zond"); a life-support research program with the announced goal of maximum closed-loop recycling using biological systems; a heavy-lift launch vehicle bigger than the now-defunct Saturn-V; an orbit-to-orbit tug, currently chemical-powered, potentially nuclear-powered. Using these blocks separately, you have space stations, space probes, perhaps even military space systems; put them all together, they spell man-to-Mars.

The Salyut-6, for example, is primarily part of a development program leading to permanently occupied, low earth orbit space outposts in the very near future. In the first four years of its operations the twenty-ton space station was occupied about half of the time by two-man teams of cosmonauts; the longest stay was about six months. A dozen robot freighters brought up supplies and replenishment of rocket propellant needed for altitude control and orbit boosting. The total weight of these supplies ultimately exceeded the original weight of the Salyut module. Cosmonaut General Georgiy Beregovoy evaluated the significance of such long-duration space missions in this way: "Their successful development is creating the necessary conditions for manned interplanetary flights." There is no reason to doubt his words.

Cosmonauts returned to Earth aboard three-ton Soyuz command modules, the heat shields of which are scoped to survive atmospheric entry at 25,000 fps. Between 1967 and 1970, however, the Soviets flight-tested Soyuz-class vehicles with reinforced heat shields. Four such unmanned probes, given the cover-name "Zond" to conceal their relevance to manned space systems, made successful circumlunar flights and survived reentry speeds in excess of 36,000 fps. Some additional reinforcement would be needed by the Mars-return speeds of 40,000 to 45,000 fps, but that should pose no real problem.

In 1970–1971, the Soviets *seem* to have tested, but then scrapped, an equivalent to the Apollo moon-landing Lunar Module, thus giving them additional experience with building the critical Mars Entry Module needed to take cosmonauts down to the surface of Mars, and then back into orbit.

Here's some simple arithmetic: the total weight of a Salyut plus four man-years of supplies plus an Earth-return module is thus just under 100,000 pounds. Remember that number.

Late in 1981, the Pentagon released a ninety-nine-page report on Soviet military power, which contained a small section on space technology. In it, the U.S. government for the first time ever explicitly confirmed the widespread reports that the Soviet "super booster," which failed miserably during a test program between 1969 and 1972, was being resurrected after a decade of redesign work. Some observers believe that the vehicle's liftoff thrust will be twice that of the U.S. Saturn-V (14 million versus 7.5 million pounds), but the only official numbers in the Pentagon report referred to the payload in orbit: "six or seven times that of the Space Shuttle." That works out to about 400,000 pounds in low earth orbit, which with the addition of conventional chemically powered upper stages to obtain hyperbolic escape velocity, translates to almost 150,000 pounds to the moon or alternately to more than 100,000 pounds to Mars.

A single launching of this booster (which is expected to become operational in the middle of the decade) could therefore send two men on a two-year fly-by mission to Mars and back. The kind of hardware that would be needed has already been flight-tested, except for the big booster.

Since a Mars-orbit vehicle would weigh at least twice as much as a fly-by vehicle, and a lander could weigh several times more, the direct approach is to launch two or more super-boosters and link their propulsive stages together in low earth orbit. A more intelligent and economic strategy is to double or triple the efficiency of the space propulsion stage. In any case, the Soviets are not afraid to deal with very heavy weights in orbit, since for the last several years they have been launching more than one million pounds of payloads into orbit every year, the equivalent of several heavy man-to-Mars vehicles.

Along the lines of greater propulsion efficiency, a recent statement by space engineer Oleg Belotserkovskiy assumes particular significance. Belotserkovskiy, who is head of one of the academic "think-tanks" closely linked to Soviet space science and engineering, would not likely be allowed to make idle, empty boasts in public, but he told a Russian journalist that the most logical next step in Soviet space propulsion was the development of *upper stages driven by nuclear power.*

Such an approach, if in fact it does reflect official Soviet thinking, makes very good sense from their point of view. Soviet nuclear engineers are certainly quite capable of creating a twenty-ton nuclear propulsion stage (perhaps to be attached to a twenty-ton propellant tank, with both units launched atop the already operational 'Proton' booster) with a specific impulse or efficiency factor of 800 to 1200. Storable chemical propellants may reach 300, and liquid hydrogen—a technology the Soviets may have decided to leapfrog—gives only about 450. The nuclear stage would be twice as efficient as any in the American inventory, or any we expect to build in the next twenty years, and Soviet policymakers are not subject to the anti-nuclear hysteria so prevalent in the West. Nuclear power, and possibly propulsion as well, has a firm place in Soviet space hardware.

From bitter experience, the Soviets know what comes of provoking the Americans into a "space race." The last time it happened, America left the Soviets in its dust within five to ten years, and an overstrained Soviet space industry suffered a string of disasters and fatalities that took a decade from which to recover. The last thing the Soviet space program now wants is an alerted, aroused America and a Congress willing once more to give NASA a book of blank checks. Thus it is unlikely that any Soviet "man-to-Mars" program will achieve public acknowledgment beyond general far-off promises of boasting cosmonauts and scientists. The absence of such a declaration cannot therefore be construed to imply the absence of such an intention.

Better diagnostic methods would include a careful watch of the kinds of space operations undertaken in the coming years. If by the mid- to late 1980s we see this "super booster," plus a year-long manned orbital endurance run, plus an advanced "space tug" (possibly nuclear, or at least fueled by liquid hydrogen), plus renewed unmanned probes to Mars, the arithmetic of their synergistic combination will be compelling: *they really mean to send men to Mars.* And by the time we recognize this, they'll only be a few years away from doing it.

11

The Home Front:
Political/Social Issues

MARS CONTINUES TO glitter brilliantly in the night sky, but
its hold on human imagination has recently suffered a hopefully
temporary eclipse. The last federal government official to publicly
advocate sending astronauts to Mars was Spiro Agnew in 1969, and
that was a bad omen. Since then, the public sentiment of flying to
Mars has seemed to shift significantly. Where once the most popular books and movies on the subject described heroic expeditions
into the unknown, the most notable Hollywood Mars movie of the
1970s, *Capricorn One*, depicted a sordid cosmic hoax and the
ultimate failure of space technology.

The U.S. government became extremely negative toward the
concept. It's worth repeating this comment which came from Jimmy
Carter's science advisor, Dr. Frank Press, who in late 1978 quipped,
"Nobody in Congress or the Federal Government or the public
has put forward a case for a U.S. manned Mars mission—and if
the Soviets decide to spend 70 billion to land men on Mars in five
years, we say 'God bless them'."

There have been voices in the wilderness. Besides the Mars Underground organized at Boulder, Colorado, a few men have urged a resumption of forward motion into space. Senator Harrison "Jack" Schmitt, who walked on the moon during the Apollo program, has published probably the most detailed and far-reaching new plan. In speeches across the country, he has revealed his vision. "In the first decade of the twenty-first century," he writes, "a space policy for our civilization should initiate a second solar system exploration decade. Bases and settlements on the moon, missions of exploration to Mars and Venus, and the beginnings of the establishment of a martian settlement, all are what these dreams are made of. The parents of the first martians are looking over our shoulders as they work and dream their way through elementary school, high school, and college. They dream of taking part in this next great expansion of the human race and its civilization."

One of the most significant philosophical discussions of the role of exploration in human civilizations was developed a few years ago by novelist James Michener, who explained, "My own life has been spent chronicling the rise and fall of human systems, and I am convinced we are all terribly vulnerable." But if there is one lesson which Michener has extracted from his historical research, it is the role of a culture's faith in the future in actually fulfilling that future.

> I also believe that there are moments in history when challenges occur of such a compelling nature that to miss them is to miss the whole meaning of an epoch. Space is such a challenge. It is the kind of challenge William Shakespeare sensed nearly 400 years ago when he wrote:
> There is a tide in the affairs of men,
> Which, taken at the flood, leads on to fortune;
> Omitted, all the voyage of their life
> Is bound in shallows and in miseries.
> On such a full sea are we now afloat,
> And we must take the current where it serves,
> Or lose our ventures.
> We risk great peril if we kill off this spirit of adventure, for we cannot predict how and in what seemingly unrelated fields it will manifest itself. A nation that loses its forward thrust is in danger, and one of the most effective ways to retain that thrust is

to keep exploring possibilities. The sense of exploration is intimately bound up with human resolve, and for a nation to believe that it is still commited to forward motion is to ensure its continuance.

Therefore we should be most careful about retreating from the specific challenge of our age. We should be reluctant to turn our back upon the frontier of the epoch. Space is indifferent to what we do; it has no feeling, no design, no interest in whether we grapple with it or not. But we cannot be indifferent to space, because the grand, slow march of our intelligence has brought us, in our generation, to a point from which we can explore and understand and utilize it. To turn back now would be to deny our history, our capabilities.

Summarizing his conclusion, Michener ended by asserting that "All the thoughts of men are interlocked, and success in one area produces unforeseen successes in others. It is for this reason that a nation like ours is obligated to pursue its adventure in space. . . ."

There is another perspective on this philosophical issue, and it is an international one. Jason Klassi, a Los Angeles space enthusiast and participant in one of the man-to-Mars studies done privately in the late 1970s, put it this way: "The challenge of the new frontier is not reserved for Americans, but is there for anyone who will accept it. If we do not take hold of the honorable opportunity to lead the way, some other nation will. Once the first step toward this enterprise is taken . . . no further step will be as great. Ceasing to push against the farthest frontier is rejecting the deepest human instinct—the striving for knowledge."

In the past, civilizations have been known to deliberately turn their back on such challenges and retreat into cultural isolation and stagnation. One of the best examples of this is the history of Ming China's abdication of a technological and cultural lead over other civilizations (including the barbarian provinces of Europe). Arthur Kantrowitz recently retold the story, but it has been around a long time (and was a favorite metaphor of Wernher von Braun's).

Early in the fifteenth century, great Chinese "treasure ships" with crews of 500 sailors were crisscrossing the Indian Ocean, visiting Bengal, Ceylon, and the East Coast of Africa. This naval power enhanced Chinese overseas influence and enriched the

Chinese economy. It also invigorated the technological progress and cultural life of the nation. But the social instabilities that such activities were bound to unleash displeased the traditionalist Confucian imperial bureaucracy, which tried to eliminate such ferment by cutting off the expeditions of the great fleets.

In 1436, the Cheng-t'ung Emperor came to the throne, and an edict was issued that forebade the building of high seas ships. Admiral Cheng Ho's naval forces were restrained and gradually reduced, and within half a century there were imperial regulations making it a capital offense to build sea-going ships with more than two masts. China, Kantrowitz concluded, "lost its leadership in technology and isolated itself as long as it could from the explosive growth of Europe."

Historian Joseph Needham, who wrote *Science and Civilization in China* (Cambridge, 1971), explained it this way: "The Grand Fleet of Treasure Ships swallowed up funds which, in the view of the right-thinking bureaucrats, would be much better spent on water-conservancy projects for the farmers' needs, or in agrarian financing, 'ever-normal granaries', and the like." Kantrowitz argued that this is the traditional mode of thought of all bureaucrats, whether fifteenth-century Chinese or twentieth-century North American: they choose safe, risk free, easily predictable policies which gradually lose ground but only slowly, generation by generation. Space exploration is seen as extravagant, but Kantrowitz asserts just the opposite. "I think the U.S. retreat from adventurous technology (unless it is soon corrected) will be regarded as the *greatest* extravagance of all such historical extravangances. . . . Leadership will not be regained by safe, pedestrian improvements alone. It will require in the future as it has in the past the bold, adventurous leaps in which risk and uncertainty are accepted. The large-scale utilization of space presents an opportunity for such a leap."

Von Braun drew similar lessons from the Ming Navy story. At a keynote address at the Goddard Memorial Symposium in 1972, he said, "Let us hope we can learn from history." There were, he claimed, "disquieting signs of an 'inward turning' among Americans, which we can only hope is but temporary. Carried to the extremes that occurred in China of the Middle Ages (this) could be nothing less than a catastrophe for modern America." He urged Americans to instead remember the value of seeking "far horizons not for pragmatic purposes alone, but for the wonder and beauty which feed the imagination and stir the soul. Nevertheless, the opening of new frontiers has always proved to be in man's vital interests, either for survival or for an improved life."

A public opinion poll in late 1981 asked the question, "Do you think that people will someday colonize the moon and other planets?" The responses were 48 percent "yes" and 52 percent "no". Interpretations of these results probably depends on the analyst's initial prejudices, but space proponents probably had the greater reasons to be pleased: even in an era *without* pro-space national leadership, a full *half* of the adult population already felt that other planets such as Mars could someday become homes for settlers. (While not part of the question, it is probably safe to assume that the majority of positive respondents felt that it would be Americans who would carry out this settlement.)

The question facing man-to-Mars proponents is how to mobilize and exploit these pre-existing sentiments in order to instigate an officially sanctioned Mars exploration-settlement movement. This needs to be done on such a scale that Congress will consistently vote for the billions of dollars needed, regardless of any one public official's term in office. That was the topic of one of the more free-wheeling and imaginative workshops at the Boulder conference: "Political and Social Issues." It was co-chaired by "BJ" Bluth and myself, with help from "Buzz" Aldrin and Leonard David, and with active participation of about a dozen other enthusiasts and would-be public opinion manipulators.

One key concept was *inertia*. If people think that man-to-Mars is a project that must be started up and laboriously set into motion, then ordinary momentum will oppose this initiation of the project at every step. Better, the workshop decided, to try to give the impression that the thrust towards Mars is already in progress and Americans would have to choose *not* to participate. That is, the options are to keep going on a space development path whose natural outcome is manned settlement of Mars, or else change direction and consciously abdicate America's leading role in space development. Such a negative decision, the conferees felt, would be a repudiation of traditional American values and practices, and the public and politicians would have to become convinced of that.

Another public opinion tactic, derived from the environmentalist movement (whose grass-roots strengths are much envied by space enthusiasts), is to avoid incurring many major opponents of the program. Potential negative opinion-makers should be identified early and neutralized with persuasion and pressure, so that the pro-Mars groups could appear to represent a consensus view.

Primarily, though, the lesson was this: national leadership must learn (or be taught) that pro-space policies are popular policies, are vote-getting policies, are policies which transform mere politicians into far-sighted statesmen whose actions go down in history. If that can be done, the money for Mars should be forthcoming.

In an engineering study designed to forecast space technology in the 2000 to 2020 time period, a team of systems analysts from the American Institute of Aeronautics and Astronautics tried to put a man-to-Mars effort in its social perspective. First, they considered the technological point of view. "A manned Mars mission would strain the propulsion capabilities in the early years of the next century, but judicious use of aerodynamic braking at Mars, plus in-situ propellant manufacturing using water (from the polar caps or subsurface deposits) would substantially ease the propulsion

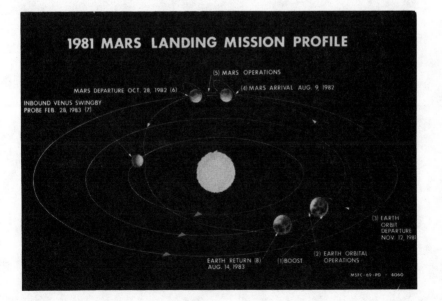

requirement. High capability nuclear systems would probably be used for advanced missions late in the twenty-year period." So far so good. That's right in line with the conclusions set down here.

But how likely is it ever to really happen? Here, the engineers turned away from gadgets and addressed human themes.

> The scenarios represent 'prudent extrapolation' of current situations and trends and are perhaps colored by the difficulties experienced by the space program in recent years. In the 30s and 40s many thoughtful and knowledgeable planners felt that manned space flight lay no closer than the turn of the century. This also was a prudent extrapolation and, had things progressed in a logical manner, might well have been accurate, perhaps optimistic. Few, if any, perceived the competitive pressure and technology jumps which both caused and allowed the tremendous spurt in space exploration that occurred. Although currently in the doldrums (which may in some degree be a natural reaction to such rapid advance) the possibility of a resurgence of mankind's desire to explore cannot be overlooked. Given free rein, man's exploratory spirit could drive him far into space by the first decades of the next century. Extensive manned and automated activity on the Moon would be commonplace and a presence, possibly permanent, on Mars would be reasonable to expect. . . .

We want Mars to be like the Earth. There is a very deep-seated desire to find another place where we can make another start, that somehow could be habitable. . . . It has been very, very hard to face up to the facts, which have been emerging for some time, that indicate it really isn't that way, that it is just wishful thinking.

Bruce Murray

12

Colonization

THE IMAGE OF the inhospitality, even vicious hostility, of Mars was placed into perspective at Boulder by a comment from Benton Clark. However bleak things looked, Clark reminded his audience, there was at least one species which definitely *could* survive under martian surface conditions. That species, Clark announced, was "homo technologicus"—us.

But the Boulder conference was interested in far more than mere survival on Mars. In fact, the atmosphere was at times downright imperialistic. Stan Kent, in referring to the first manned visit, which need not depend on on-site propellant manufacture, used the military-sounding name of "the beachhead mission" in describing it. Jim Cutts stressed the development of surface science, but urged that it be concentrated on the establishment of an enduring presence, not just a campsite. "Our basic goal should be to learn how to extract resources, for constructional materials and consumables," he remarked. "All other scientific research is secondary." And in summarizing the conclusions of the mission strategy workshop, Jim French justified a base on the grounds that it was "for purposes of evaluating Mars as an object of long-term habitation and colonization—not just for scientific curiosity."

One of the "key concepts" to come out of the conference (about twenty were listed in a summary document published by the Mars Underground officials) was to expand the base as rapidly as possible and as independently as possible. "The goal must be autonomy," the summary document stressed, repeating French's directive that scientific research be concentrated on the question of human colonization and habitation.

Richard Johnson, who like many other NASA employees came to the conference out of personal interest and in the face of official concern that even the *appearance* of official support might possibly be misconstrued by the presence of any NASA people, revealed that a small study on human colonization of Mars had in fact been conducted at the Ames Research Center in 1980 and 1981. Entitled "Mars Settlement Program—A Program for Solar System Exploration which is Focused on Mars Settlement," the study (not hitherto publicly discussed, according to Johnson) explored the possibility of the establishment of a human Mars colony with about 100 people, largely self-reliant and permanent, in the mid to late twenty-first century (about 100 years from now). Further exploration of the solar system would be supported by, and staged from the Mars base.

The Ames-Mars study listed four primary program justifications: scientific, resources, habitability, and stepping-stone concepts. The purpose would be to unify and give direction to a NASA-wide program goal.

The scientific questions deal with the origin of the solar system, comparative planetology, and dynamic solar system processes (such as variations in solar output) that affect both Earth and Mars. Long baseline interferometry, a technique in radio astronomy that utilizes widely separated receivers to vastly improve the resolution of distant cosmic observation targets, was highlighted as an early potential high-return scientific activity for a Mars base. Other questions in geology, geophysics, and exo-biology were also listed.

The resources questions deal with inventories of potentially useful materials, along with the development of processes for extraction, refining, and fabrication of desired products. (The ma-

MARS SETTLEMENT PROGRAM
HUMAN SPACEFLIGHT/OPERATIONS

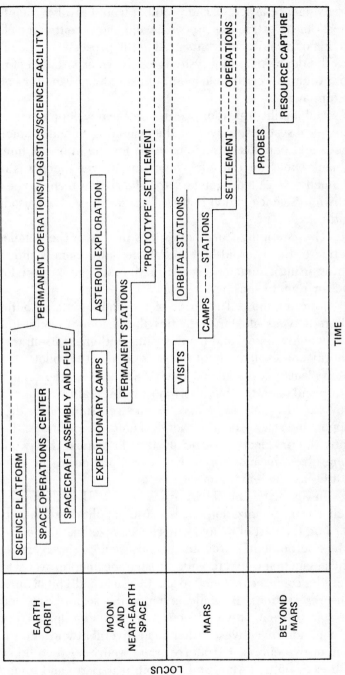

The NASA-Ames plan for martian settlement, unveiled at the Boulder conference in 1981.

terials science aspect of the problem would involve such technical questions as extraction processes and the potential for closed recycling of chemical reagents used in the processes.) Also of interest was the development of systems for the transportation of materials from their source to their point of use, half a planet or half a solar system away.

Habitability questions center on the development of synthetic ecological systems, as well as the evaluation and assessment of the needs and benefits of such an effort for the future of humanity—a blank check for blue-sky theorizing if ever there was one. Additionally, so-called planetary transformation technologies, better known to science fiction buffs as "terraforming," was also listed for study.

The "stepping-stone" questions deal with the detailed examination of the asteroid belt and of the outer planets and their satellites, using a human presence on Mars as support, both for probes and for remote sensing.

Johnson showed a series of viewgraphs which outlined the different types of Mars activities that colonists would undertake: they would support and maintain the settlement itself; they would identify and evaluate local resources and set up pilot-scale exploitation plants; they would conduct Mars-directed science; they would set up and operate a radio astronomy observatory on or near Mars; they would promote the settlement's self-sufficiency; they would support the staging of off-planet exploratory expeditions; they would study the prospects of planetary transformation; they would study themselves, by monitoring human adaptation to the remote and hostile martian environment.

In summary, the NASA Ames-Mars 1981 study suggested that such a Mars colonization (or, because of political sensitivity to that word on the part of many formerly colonized nations of the world, "Mars settlement") program would accomplish several ends. Its official sanction would provide a sustaining long-term goal for NASA (Boulder conferees seemed to feel this smacked a bit of bureaucratic self-preservation). It would enhance the scientific exploration of the solar system. It would focus manned and unmanned programs on common objectives, rather than divergent and often mutually competitive efforts. Extraterrestrial resources would be evaluated and used, both at Mars and in orbit near Earth and someday per-

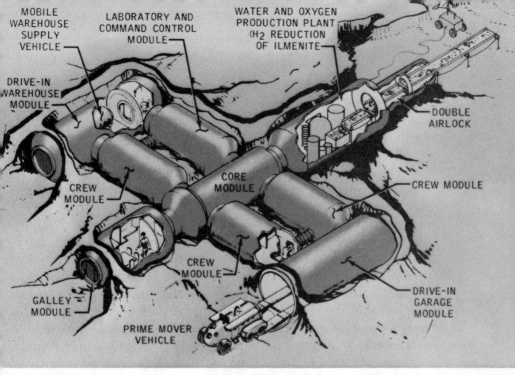

MOBILE WAREHOUSE SUPPLY VEHICLE

LABORATORY AND COMMAND CONTROL MODULE

WATER AND OXYGEN PRODUCTION PLANT (H$_2$ REDUCTION OF ILMENITE)

DRIVE-IN WAREHOUSE MODULE

DOUBLE AIRLOCK

CREW MODULE

CORE MODULE

CREW MODULE

CREW MODULE

GALLEY MODULE

DRIVE-IN GARAGE MODULE

PRIME MOVER VEHICLE

The first permanent Mars surface base will probably look a lot like this.

haps on Earth's surface. Lastly, Johnson proposed, the program had the potential to receive and maintain public interest and support over the very long development and build-up period.

One of the early detailed plans for a build-up of the initial martian base was drawn up by spaceflight visionary and meticulous engineer Krafft Ehricke. By 1970 he had already published numerous lengthy studies which called for long-duration, semi-permanent bases from the very first manned surface mission.

In fact, Ehricke's plan did not even include an Earth-return vehicle on the first landing expedition. His astronauts would spend a full synodic period (about two years) on Mars, and would then be picked up by a second expedition the sole purpose of which was crew exchange and return to Earth. Although two launches were needed instead of one, Ehricke argued cogently that such a plan was *safer* (no need for the critical Mars ascent stage to endure more than a year's exposure on the potentially dangerous surface), offered greater mission payload (without a return stage, the Mars spacecraft could be twice as heavy), and spread out the technological challenges and peak expense over two windows rather than just one.

The time at the planet for the first crew would be about 750 days, and the only disadvantage from the crew's point of view is a longer total mission time—up to, or more than three years away from Earth.

The next stage, per Ehricke's plan, would involve expanding the pickup mission into a full relief expedition. Meanwhile, one-way supply delivery vehicles weighing a million pounds in Earth parking orbit, and using chemical propulsion, can carry up to 240,000 pounds into Mars orbit. A nuclear engine system of the same weight could deliver more than 300,000 pounds of cargo and on-site re-fuelling could send these vehicles back to Earth. Not mentioned by Ehricke, but described later by the Stevenson and Staehle studies, is that a solar sail or solar electric system of the same initial weight could deliver twice as much payload for a twice as long voyage (for a longterm buildup, that sacrifice in speed is probably well worth it). Meanwhile, Ehricke also envisaged a round-trip personnel module of 100,000 pounds, which could be carried by another million-pound launch vehicle using combined nuclear and solar-electric ion propulsion. Between them, these designs could support a step-by-step buildup of the Mars landing into a Mars settlement.

The second mission of Ehricke's plan would exchange the crew, and Ehricke suggested an even dozen astronauts. The third mission brings eighteen crewmembers, new supplies, two new landing craft, two reusable orbit-to-surface tugs, and a large de-ployable shelter for setup on the surface. Mission Four brings twenty-four crewmembers, an additional shuttle vehicle, and equipment for operational resource exploitation. By the fifth ex-pedition, ten years into the sequence, crewmembers have the op-tion of staying over on Mars for an additional hitch; these people would have been away from Earth for more than five years when they finally got home.

Two types of events would occur in this time frame, give or take a few years, which would mark the humanization of the planet: deaths and births. The implications of both, and the back-home public's reaction to both will go a long way toward determining the ultimate destiny of the fledgling Mars settlement.

The economics of safety are grim. Budget officials may be able to prove that for the same cost needed to send ten astronauts and guarantee 99.95 percent safety, twenty could be sent with only 90 percent safety, implying that one or two or three of the astronauts would die on Mars. From the standpoint of positive results, either ten could make the trip, or maybe seventeen or eighteen or nineteen; the *dollar* cost is the same but the human cost is different. Are the things that the extra astronauts could do worth the additional human cost? Many would argue so, and most vocal among them would probably be the astronauts themselves. People, highly talented, stable, dedicated, and rational people, have historically always been around, who are willing to go to new lands under such conditions.

Any major Mars exploration-exploitation effort must be able to accommodate the deaths of individuals, the abort of occasional missions, and perhaps even the loss of an entire expedition's personnel. The last situation could be particularly painful in plausible situations where doom is absolutely certain via laws of life-support consumables or orbital mechanics, even while the crew has weeks or months yet to live. They almost certainly would continue with whatever mission phases were feasible, while debriefing and preparing records on the lessons learned in their experience, all in the full, agonizing glare of world-wide attention. And the next time, there would still be willing volunteers.

It is interesting in this light to note how Soviet space commentators have rationalized their spaceflight fatalities (the first Soyuz in 1967 and the first Salyut mission in 1971 both resulted in the deaths of the entire crews), to the point where Russian citizens are positively *proud* of the proof that their 'boys' are pioneering the farthest reaches of the most dangerous frontier, and have been willing to take the risks and pay the ultimate price to bring honor to the motherland. All previous frontiers have exacted such payments in blood, and that burden has always been accepted as unavoidable.

The happier side of the coin is the possibility that human beings will be born on Mars within a few years of the first semipermanent settlements. Such births, which in the light of current biological knowledge will almost certainly be planned (if not officially approved!), will imply the intention of the parents to remain on the planet for a significant length of time. Unlike the case of Antarctica, where the first human baby was born in 1977 (of Argentine parents, at a small base set up to support official Argentine territorial claims), transportation may not be routine enough to support the medical evacuation of infants back home. The child would have to be fairly mature, say ten years old or more, before the trip back would be feasible. Also, the decision to stay at Mars (on Mars, in Mars, under Mars) for the rest of one's professional career will be a momentous one, and he or she will be the first true colonist of the planet. From that, it's only a small step to deciding never to come home at all. People may be making that decision within the second decade of human activity around Mars.

A biological perspective on the colonization of Mars was offered at Boulder by Dr. Bassett Maguire of the University of Texas.

Artist's conception of life in a martian settlement.

The silver-bewhiskered Maguire entranced the listeners with a classroom style combining remarkable insights with a dry, subtle wit, clearly the type of crewmember to bring along on the Mars expedition. An excerpt from his prepared paper follows.

> It may very well be that historians of the far future will view the establishment of self-sufficient, extraterrestrial communities of plants and animals (including humans) of Earth origin to be at least the second most important event in all the long continuum of the living organisms of our planet. Only the origin of life on Earth may be considered of greater significance (and perhaps not even that). . . . Species on Earth, if they are to persist for long, and are to become what biologists call highly successful, must undergo considerable expansion of their ranges after their origins. Species which do not extend their ranges from their points of origin persist for only relatively short periods before they become extinct. . . . The relatively uncommon spread of species from their planets of origin may be more significant than the independent origin of life on different planets.

The use of martian gardens for food crops, and the growing sophistication of regenerative closed-loop recycle systems for life support, is bound to lead to the growth of small closed biological systems on Mars. Maguire addressed the challenge of such a "microcosm" at Boulder, referring particularly to those with relatively small volumes and with only small subsets of the total biota of Earth.

"Until one stops to consider the matter," he wrote, "one generally does not recognize how many of the needs of plants and animals are 'automatically' taken care of by the Earth's normal dynamics. For example, closed greenhouse-like systems will run out of photosynthetically usable levels of carbon dioxide within a few tens of minutes when they contain dense populations of plants carrying on photosynthesis." One possible solution, based on his own laboratory work, is that "an 'ocean' containing an appropriate mixture of potassium carbonate and bicarbonate may serve as an aid in buffering the carbon dioxide concentration of the small volume atmosphere of such systems." Maguire also nominated *pineapples* for early martian gardens ". . . because they have 'Crassulacean Acid Metabolism' which allows them to take up carbon dioxide from the atmosphere at night and to store it for photosynthetic use during the next lighted period."

Up until now (1982), attempts to create self-sustaining microcosms have not met with notable success. Maguire described "fairly rapid dissolution of the communities and death of the systems. . . . We do not yet know what the critical factors were which were responsible for the collapse of the ecosystems."

Designing a Mars mini-ecology will be the subject of research for decades, but Maguire gave this preliminary assessment. "A recently suggested list of plants for incorporation into closed ecological life-support systems includes (only) soybean, sugar beet, wheat, rice, and lettuce. Our highest yield crop—corn—can at

Larger settlements might take the form of domed cities or even burrow underground.

once be added to this list for a solid body such as Mars, where area rather than volume will probably be the important determining factor. . . . It seems to me that this list needs to be further enlarged to include possibly onions, herbs, potatoes, yams, amaranth, and others. Because half or more of each plant will be indigestible by humans it probably will be profitable to include appropriate cellulose digesting animals. Not only will these provide for high quality protein and (not a trivial point) for gustatory satisfaction, but they will make the eating of an adequately balanced diet easier. . . ."

These major colonization activities will put a strain on the transportation systems which were barely sufficient to support initial expeditions. Within ten to twenty years of the first expeditions, if expansion of the human presence is to continue at a reasonable rate, the technology of interplanetary transportation must be revolutionized.

Krafft Ehricke put it this way. "For manned interplanetary flight above bare minimal missions to Venus or Mars, high specific impulse is more than desirable; it is necessary. . . . The many attempts to justify the use of marginal and submarginal nuclear and chemical propulsion systems for manned interplanetary missions in the 80s or even later are difficult to understand."

The interplanetary engines of the middle of the twenty-first century—or even sooner, with unpredictable but traditional breakthroughs—could be based on high-yield nuclear energy engineering which yields efficiencies five to ten times as great as the promised value of the old-style NERVA solid-core nuclear stage of the 1960s (itself twice as efficient as today's best chemical rockets). Conceptually, several different types of nuclear engines can be described even today. The difficulty is in finding ways to control the reactions and in constructing material assemblages able to contain such fury.

One frequently mentioned device is the "gas core nuclear reactor," so named because the nuclear fuel is reacting at temperatures high enough to melt any candidate core materials. The great heat translates into the ejection velocity of the propellant, and ejection velocity is directly proportional to efficiency. In a gas core system, the nuclear material must still be brought in close contact with propellant, heating it and expelling it violently while not losing any nuclear fuel in the process, so it can be recycled. Various bizarre schemes were studied in the 1960s, including one in which a stream of fissioning nuclear material is surrounded by a sheath of hydrogen propellant: the nuclear material is captured in a scoop and recirculated, while the super-heated gas is released through a thrust nozzle and new gas is injected into the chamber. Physical contact must be as close as possible without actual mixing that could cause nuclear fuel loss.

ORION-class pulse-boost systems have been discussed since the early 1960s, and everyone seems to agree that they are extremely feasible and totally out of the question for political reasons—since they actually use a continuous string of nuclear detonations behind the spacecraft, buffered by a massive spring-mounted "pusher plate" which ablates a few fractions of an inch with each blast. The minimum size for an ORION man-to-Mars ship would be about 200,000 pounds, of which 50,000 pounds is payload; roundtrip time would be as little as four months plus stopover. Specific impulses of between 4,000 and 10,000 seconds seem attainable if and when the international political and environmental situation allows it. An argument *for* such a system is that it would be an extremely beneficial use of thermonuclear bombs! (There is also a "fusion micro-bomb" design which could overcome political problems since its laser-induced pulses are only in the range of tons of TNT.)

Fusion power could also become available for spacecraft propulsion, promising medium to high thrust for days on end with efficiencies of 10,000 to 100,000 seconds. One proposed design, the rigatron engine of Dr. Robert Bussard, could, according to its proponents, take a 2 million-pound spaceship from Earth to Mars and back with one fueling in a month, with a payload of half a million pounds. A crash program could see such a vehicle built within twenty years, say its enthusiasts.

MAMA is another option—"matter-antimatter annihilation", or the famed "antimatter drive" of *Star Trek* fame. Presuming that a reasonable supply of antimatter could be manufactured and stored (problems people are only beginning to appreciate), spacecraft with such engines could range the solar system. Simple calculations show that a vehicle maintaining an acceleration equivalent to one full G (Earth surface gravity, about thirty-two feet per second) could reach Mars in as little as two days, or reach Pluto in two weeks. The potentials, and the problems, belong to the next century!

The synthesis of our ideas has resulted in the formulation of a master plan, a sort of recipe for terraforming, which uses technology to begin the process and then relies increasingly on biology to maintain and stabilize it. According to the Gaia Hypothesis, planets are controlled by their biology. What we propose for Mars is to give this biology a beginning, testing our hypothesis, and reaping the human rewards of terraforming.

> *Penelope Boston, director of the "Mars Project" at the University of Colorado's Laboratory on Atmospheric and Space Physics (1979)*

13

Terraforming

IN THE SUMMER of 1975 a group of scientists met near San Francisco to discuss Mars. The Viking probes were about to be launched, and their goal was to measure the conditions on the surface of the planet. The scientists had a far more long-range goal. They were to discuss whether or not Mars might be *habitable* by any form of Earth life, or, if it were not, *whether it could be made so.*

The study, sponsored by the nearby NASA Ames Research Center, came to tentative positive conclusions on both questions. Certain types of terrestrial algae and lichens might be able to survive on the martian surface, particularly if they could be bioengineered to enhance characteristics useful to withstand the cruel conditions there. Additionally, the entire climate of the planet could be altered through the spread of these Earth-derived plants in a period of several thousand years; if additional efforts were made via engineering and technological means, that period could be reduced to thousands of years or even less.

Such a concept is called "terraforming," or "planetary engineering," and it refers to the wholesale rearrangement of a planet's environment by modifications of its energy balance or of its material composition. Frequently such modification can be effected by attacking chinks in the ecological armor of the planet, "pressure points" at which small changes will have disproportionate impact on the whole planet's climate machine. The identification of such "pressure points" and of the technologies which can be used against them will be a significant science in coming centuries.

For Mars, a goal would be to provide the planet with thicker, breathable air, at temperatures high enough for lightly protected human beings to venture forth on its surface. Plant and animal life would spread across the now-barren surface, tinging the red rocks with green, and turning the red sky into a beautiful dark blue. Liquid water would flow again on the surface, and the eons-dry channels and gullies would become wet with new rains. Rainbows would appear in the sky, symbolizing not the restraint of floodwaters (as in *Genesis* after the Flood) but their *release* from eons of imprisonment within the permafrost.

Several strategies are applicable to Mars when the decision is made, centuries from now, to reconstruct it in Earth's image. Thermal manipulation, via both giant mirrors in space and alterations of the albedo, is a leading tool for would-be terraformers. Material manipulations, in particular the addition of large quantities of atmospheric nitrogen, can also occur. Lastly, specific techniques for Mars could involve the more restricted goal of merely terraforming a few corners of the planet at first, rather than the whole world.

The thermal balance of localized regions on Mars can be influenced by pouring energy in (using mirrors in space) or by absorbing more solar energy which happens to fall (by darkening the surface). Giant mirrors, many miles on a side but thinner than aluminum foil, have long been proposed for use around Earth, for various applications such as power generation, lighting, local climate control, hurricane steering, and other uses. Man-made modifications of Earth's surface albedo have been going on for millenia, mostly accidentally affecting rainfall patterns rather than temperatures. In both cases, the technology of terraforming will be well understood and mastered at Earth long before they are applied to Mars.

Phobos and Deimos have roles to play in this process. Their carbonaceous chondrite composition can provide both the very dark material needed to increase the sunlight absorption of areas on the surface (transported from the moonlets by a small aimable electromagnetic cannon), as well as the metal ores such as magnesium for the local manufacture of mirrors (unless they are built more cheaply near Earth and then "sailed" by sunlight pressure out to Mars orbit). Needed control points on one or both of the moonlets will require a permanent human presence there, probably burrowed deep in their interiors.

The targets for this thermal manipulation include the ice caps, the hypothetical Tharsis glaciers, and other areas of rich permafrost deposits. The purpose would be to melt and evaporate the trapped volatiles, thus increasing the pressure of the martian atmosphere.

A thicker atmosphere has numerous climatic advantages. Most obvious is the enhanced "greenhouse effect" by which incoming solar radiation is trapped in the lower atmosphere. Equally valuable is the growth of the role of winds in the horizontal distribution of tropical heat to polar regions (heat is not carried well by the current thin martian winds). Both effects contribute to the continued maintenance and enhancement of the warming trend set off by these artificial modifications of the planet's thermal balance.

Reasonable computations for the rate of warming show that the surface conditions on Mars could stabilize at a much thicker, warmer atmosphere after a period of about 1,000 years, maybe half as much, maybe twice as much. So great is the thermal inertia of the surface of Mars that substantial acceleration of this schedule would require unreasonably high energy input and absorption rates. No shortcuts are readily apparent under currently conceivable technologies, but we have a century or two to get smarter.

That is, there do not seem to be any shortcuts to making the *whole* surface habitable. However, there *are* shortcuts for creating small habitable *oases* by technologies which also help in the slow planet-wide warming process.

One way is to drop rocks on Mars, from the asteroid belt or

as far away as Saturn. Far-ranging explorers in the late twenty-first century could locate volatile-rich bodies, on the order of tens of miles across. The objective would be to alter their orbits to lead to a collision course with Mars, causing a tremendous impact explosion which would gouge a miles-deep crater on the surface. This crater would be warm because of the energy of impact, and it would accumulate higher-than-average air pressure because of the new low elevation of the interior surface. Such an oasis could be an early habitat for Earth life on the open surface of Mars.

Getting an icy asteroid ("ice-teroid"?) to Mars would certainly not be a trivial process! Its orbit could be nudged by setting off thermonuclear blasts on one side, which would throw off debris and consequently shove the rest of the asteroid in the other direction. A series of such blasts would use up a significant fraction of the object's material, but the residue would be slowly eased into a new orbital path. Final trimming of the new orbit could be accomplished by installing several solar-powered electromagnetic cannon on the asteroid to expel streams of debris, thus delivering a small but continuous thrust force.

Steering the asteroid past Jupiter would be useful. This would result in considerable energy and time savings, since the giant planet's gravity could then twist the trajectory onto a direct path intersecting Mars. The greater the impact energy at Mars the better (it makes a deeper hole, warmer rocks, and better pulverization and distribution of volatiles trapped in the asteroid and the martian crust). So high speed encounters are to be preferred, which is just the opposite strategy of that with manned spaceships. One possibility is to use a "reverse Jupiter swing-by maneuver" to turn the doomed asteroid onto a retrograde orbit around the sun, moving in a direction opposite to all the other planets and asteroids. This would allow a head-on rather than an overtaking collision with Mars, and could thus provide up to ten times as much impact energy.

The movement of the asteroid from its eons-old peaceful orbit to its impact on Mars could be accomplished in twenty or thirty years. Planetary motions are the major constraints on this schedule, not available energy sources. And for a centuries-long program, a few decades is not an unreasonable interval.

One such impact would not be enough. The actual mechanics

of formation of giant craters are still subject to great theoretical dispute, but the best estimates are that even a crater 100 miles or more wide is never more than a few miles deep, even during the rebound immediately after formation. To be an effective "trap" for gathering the atmosphere, an oasis-crater should be up to six or eight miles deep and at least 100 miles across. That would require several impacts onto the same point, each peeling back another mile or two of the crustal rocks.

Perhaps the incoming object could be cut up into chunks about a year before impact, with each piece targeted separately to hit the same point but several hours apart.

The first location for such a multiple-crater should be a region already naturally as low as possible, and preferably far from interesting geologic sites, which should be preserved for scientific purposes. The Hellas Basin immediately comes to mind as a leading candidate. It is a wide, featureless plain in the ancient cratered terrain of the southern hemisphere highlands, with a depression as low as any other point on the planet. It is so wide that several different oasis-craters could be constructed within its surrounding rim-wall mountains.

The end product of these catastrophic impacts would be a wide circular structure, its surface covered with hot rubble and its low center covered with water. Air would be thick enough to support human beings with simple oxygen masks. The rocks would be warm enough from the energy of the smash and from sunlight trapped in the thicker air, so that Antarctic-style clothing would be the most needed to walk around outside. Some regions, particularly near hot springs and geysers, would be even balmy.

Earth life forms could be immediately introduced onto the surface and into the "crater lake" waters. Plants would thrive under such conditions, and the entire atmospheric carbon dioxide could be transformed within a few decades. Enough oxygen would accumulate to allow animal life to survive, as long as the carbon dioxide concentrations remained low enough to keep from poisoning such creatures. Some neutral buffer gas such as argon must be present in large quantities to prevent planet-wide forest fires among the plants. Nitrogen also must exist in a significant amount to support biological cycles; if it is not freed from the crust by early micro-organism activity, it will have to be imported at great expense and time penalty from Titan.

And as temperatures elsewhere climb, the air will thicken all across the face of the planet. Oceans in the craters will fill up and inundate these first biological pioneers on Mars. Eventually, a pole-girdling ocean network could cover much of the northern hemisphere. Vast coastal areas would host a diverse imported ecology, tailored and engineered for particular martian conditions. And at the top of this biosphere hierarchy, acting as stewards and "forest rangers" for the whole blooming planet, would be the human settlers, by the thousands, and soon by the millions.

All this world is heavy with the promise of greater things,
and a day will come, one day in the unending succession of
days, when beings, beings who are now latent in our
thoughts and hidden in our loins, shall stand upon this Earth
as one stands upon a footstool, and laugh and reach out their
hands amidst the stars.

H. G. Wells, 1903

Afterword

CONQUEST OF MARS

Up until now, the most enduring human monuments have
been tombs, and the most awesome attempts at organization on a
grand scale have been wars. Space exploration has shown new
possibilities: our footprints on the moon will still be fresh when
the pyramids and even the Himalayas are dust; the society-wide,
organized industry and ingenuity which sent people to the moon
has no historical peacetime parallel, and in fact approaches in scale
the terrible dimensions of intertribal warfare on a worldwide basis.
The human instinct for destructive violence has met a match in
the field of spaceflight and the hitherto eclipsed instincts for curiosity, construction, and cooperation. Such a momentous cultural
event needs to be encouraged so its implications will be absorbed,
however slowly, throughout our planetary civilization.

The Mars endeavor offers us the opportunity to carry this
lesson to a logical and positive conclusion. While it is true that
humanity has come to the point where it is capable of reducing a
living planet (our own) to a blasted cinder, the realization is slowly

growing that the opposite process has also become possible: humanity has reached the point where it is capable of elevating a dead, barren, empty world into another thriving, fertile habitat for life.

Doing so—or even dreaming of doing so—may be a match for the darker, destructive side of the human spirit. Such a life-giving project would be a memorial and a testimony to the best human attributes, and its accomplishment would both actually and symbolically dwarf the earlier predominant succession of Earth's wars and deaths.

Despite the namesake of the mythological god of war, Mars can be remade into a symbol of life. Over the coming centuries, the blood-red planet Mars could be gradually replaced in the skies of Earth by a soft-colored, gleaming, green-tinted jewel, reflecting the spread of life across its surface. Even the planet's color would no longer be an indication of obstruction, but would truly signify a kind of celestial hope from our neighbor world. The metamorphosis of old Mars into a living, terraformed planet would be more than a metaphor of the ultimate conquest of the god Mars' influence in all terrestrial civilizations, or of the victory of life over death which has been the spark behind so much human aspiration. The spread of earth-born life beyond the world of its biological origin would be an event of galactic significance, both for what would still lie ahead of a newborn multiplanetary human civilization, and for what would be left behind.

So let us conquer Mars.

Appendix 1

The Moon & Mars

WHILE PEOPLE ARE on their way to Mars, what, if anything, will other people be doing on the moon? The regions near Earth, in the so-called low earth orbit, will certainly be busy with manned traffic, including space stations and outposts such as advanced Soviet "Salyut" and "Kosmograd" vehicles, or the long-proposed American "Space Industrial Park" (formerly the "Space Operations Center") or European and Japanese mini-stations attended by advanced manned space shuttles. Manned sorties to, and semi-permanent occupation of, the "geosynchronous orbit" (GEO) with its convenient twenty-four-hour period is also very likely. But it would be useful to know if manned lunar activities will be resumed in the same time period, in a competitive, complementary, or supportive role vis-à-vis the man-to-Mars effort.

Some sharing of technology is definitely feasible: propulsion systems, life-support systems, crew systems, avionics and sensors and mechanical systems, all could share developmental expenses. As one example, a trans-Mars propulsion stage could put a modified Mars "Mission Module" into lunar orbit to serve as a semi-permanent base, from which manned sortie vehicles (with life-support and propulsion systems built with commonality with a Mars Entry Module) could descend periodically to the lunar surface. Practically all that equipment would be common, and only the unique operational expenses (such as fuel bills, planning and training expenses, and long-distance radio-link charges) would be incurred.

The logic of returning to the moon in the mid-1990s depends on science, on technology, and on practical applications. Analysis of Apollo data has proceeded to the point where new productive exploratory strategies can be formulated. Technological capabilities will have matured to the point that renewed manned expeditions will be relatively cheap. And there are lunar resources which could be utilized on the moon, in nearby space, or for support of the man-to-Mars program.

First to be investigated is the possibility of lunar ice deposits. Because the moon's axis of rotation is nearly straight up relative to the ecliptic (the plane of its revolution around the sun), there are deep crater interiors at both poles where sunlight never falls. Calculations have shown that in the last few billion years, water ice from passing comets or from occasional lunar volcanic outbursts would have frozen out in these super-cold spots and remained stable. The ice could form layers interspersed with dirt thrown from distant meteor impacts; up to several feet of ice should have been laid down over the eons. All of this assumes that the moon's axis has not changed significantly, and there is no known reason to suspect that it has.

The ice would be detectable by a small low-flying polar orbit satellite. Such a vehicle, called "Lunar Polar Orbiter," was planned by NASA in the late 1970s but cancelled because of the budgetary restrictions during the Carter Administration. A catchier title might have helped, and "Prospector" describes its purpose, so let me nominate that name. The probe could be a close relative of other "Prospector" water-sniffers sent to orbit Mars around 1990.

The discovery of water on the moon (and it's a long shot, to be candid) would immediately raise the possibility of lunar-based support for the assembly and/or fuelling of the man-to-Mars expedition. Using solar or nuclear power, the water could be split into oxygen and hydrogen, liquefied, and stored as cryogenic propellants.

Closely connected with the concept of exporting lunar-derived resources is a remarkable machine now known as the "mass driver." It is actually a very old concept, dating back at least thirty-odd years to stories by Arthur C. Clarke, where it was called an "electromagnetic catapult," a much more descriptive title.

Essentially, buckets containing exportable material are accelerated through a sequence of rings, each of which magnetizes momentarily to pull the package forward. In recent years, experimental work by Henry Kolm at MIT and Gerard O'Neill at Princeton has advanced the "state of the art" to the point that O'Neill's "Space Studies Institute" has published a design for a moon-based horizontal solar-powered 200-G, 1,000-foot-long catapult weighing about 60,000 pounds and capable of annually propelling eighty times its own weight out into space, where special craft would retrieve and store the cargo for utilization. Generally, the typical cargo is thought of as moon ore for in-space refining, but cannisters of propellant are also plausible.

Such a fuelling operation for outward-bound spaceships could occur in low earth parking orbit, *or* out near the moon. Since it takes about three-quarters of the required trans-Mars delta-V to get into a low lunar orbit, possible savings are not immediately obvious until you realize that if the Mars ship's "trans-Earth stage" (the one to be used to get back from Mars with) were fuelled at the moon, it could save a whole lot of propellant in the launch from low earth orbit since it would be only empty tankage. In fact, in such a strategy separate trans-Mars and trans-Earth stages might not be needed at all: a fully fuelled injection stage could take the Mars spaceship out to the lunar orbit depot, refuel, and subsequently complete both the remaining trans-Mars injection burn *and* the subsequent trans-Earth burn much later. Alternately, the lunar-manufactured propellants could be sent down from the moon to the spacecraft assembly point in low parking orbit. An aerobraking system on these tankers would allow transportation across the intervening 250,000 miles at a fraction of the energy required to lift equivalent propellants up from Earth a few hundred miles below.

Launching towards Mars from lunar orbit would result in some not insignificant inefficiencies in the combinations of escape velocity and the "hyperbolic excess velocity" needed to reach Mars. This would amount to several thousand fps of required extra propulsion, which would really eat into the bonus of having the moon depot. Additionally, a few navigational problems concerning orbital plane adjustments would also be introduced and this would cost some additional propellant during the final departure.

A far more serious problem would be the basic cost of setting up such a lunar propellant factory. There's no way it could make economic sense even for a dozen every-other-year Mars expeditions, because the set-up cost would be many times as much as the traditional up-from-Earth transportation costs of propellants. A NASA study on giant Space Solar Power Stations showed that moon-based resource extraction only made economic sense after the first dozen or so giant Manhattan-sized structures were completed, and a man-to-Mars expedition is a lot smaller than a Space Solar Power Satellite.

There are several conflicting trends here. Other motivations might make lunar mining an economically viable option without any consideration of Mars flight. Giant solar power satellites (as just mentioned), or space colonies, or asteroid mining expeditions, or a military buildup, are just such possibilities. Under these conditions, the add-on cost of *extra* operations needed for the man-to-Mars spaceships would probably be very reasonable. But if such other options *were* realized in the late 1990s, they would represent large, expensive manned space programs which probably could not co-exist with the Mars exploration mission development (which could then be pushed off into the 2010–2025 time frame).

At the same time, a modest, reasonably funded manned lunar exploration program in the 1990s would not in itself be competitive with the ambitious man-to-Mars plans. Properly designed, the two parallel efforts could be mutually supportive, as described at the beginning of this chapter. Astronauts could visit the lunar surface for extended periods to emplace, service, and retrieve instruments, to gather samples both from the surface and from considerable depths via drilling, and to perform habitability and resource utilization experiments whether water ice is ever found there or not. Much of the advanced crew equipment for the Mars astronauts— spacesuits, rover-jeeps, shelters, even the landing module itself—

can be field tested by lunar astronauts. In fact, the flight crew for the Mars expedition may be chosen to consist entirely of astronauts who have already lived and survived and thrived on the moon.

So when human beings triumphantly step out onto their third world, other human beings on Earth *and* on the moon will be watching.

Appendix 2

What Can You Do?

KEEP IN TOUCH with groups now forming to coordinate private research and experimentation. The "Mars Underground" (P.O. Box 4877, Boulder, Colorado 80306) has already organized two colloquia, in Boulder and in Houston. The *Planetary Society* (110 S. Euclid, Pasadena, California 91101) is a strong advocate of new Mars probes, and has a nifty newsletter, too. "Delta Vee" is an enthusiastic and imaginative new West Coast group which has sponsored such events as the Viking Fund (they're at 3033 Moorpark, Suite 27, San Jose, California 95128). The British Interplanetary Society (27-29 South Lambeth Road, London SW8 1SZ) is beginning to sponsor symposia on manned planetary exploration; the first such meeting was held in January 1982. Another group is the World Space Foundation (P.O. Box Y, South Pasadena, California 91030).

There are other space enthusiast organizations with less relevance to man-on-Mars. They include the National Space Institute (Suite 203, 600 Maryland Avenue SW, Washington, DC 20024), the L-5 Society (1060 East Elm, Tucson, Arizona 85719), and Dr. Gerard O'Neill's "Space Studies Institute" (Box 82, Princeton, New Jersey 08540).

Bibliography

American Institute of Aeronautics and Astronautics. 1966. *Stepping stones to Mars*. New York: AIAA.

Ash, R. L.; Dowler, W. L.; and Varsi, G. 1978. Feasibility of rocket propellant production on Mars. *Acta Astronautica* 5:705–724.

Averner, M. M., and MacElroy, R. D. 1976. *On the habitability of Mars—an approach to planetary ecosynthesis*. NASA SP-414. National Technical Information Service, Springfield, VA.

Batson, R. M.; Bridges, P. M.; and Inge, J. L. 1979. *Atlas of Mars—the 1:5,000,000 map series*. Washington, DC: NASA Scientific and Technical Information Branch.

Blunck, Jurgen. 1981. *Mars and its satellites: a detailed commentary on the nomenclature*. 2nd ed. Hicksville, NY: Exposition Press.

Bluth, B. J. 1981a. Sociological and psychological factors for a Mars mission crew. Paper delivered at the Case for Mars Conference, Boulder.

Bluth, B. J. 1981b. Sociological aspects of permanent manned occupancy of space. Eleventh Intersociety Conference on Environmental Systems, Spacecraft Operations Center session, 13–15 July 1981.

Bogard, Donald D.; Duke, Michael B.; Gibson, Everett K.; Minear, John W.; Nyquist, Larry E.; and Phinney, William C. 1977. *Considerations of sample return and exploration strategy for Mars*. Lunar and Planetary Sciences Division, NASA Johnson Space Center, Houston.

206

Boston, Penelope J. 1978. Manned exploration of Mars. Mars Study Project, University of Colorado, Boulder.

————. 1979. Agricultural vascular plants for low-pressure martian greenhouses. Paper delivered at the Tenth Lunar and Planetary Science Conference, Houston.

————, editor. *The case for Mars*. San Diego: Univelt, American Astronautical Society.

Bradbury, Ray, et al. 1973. *Mars and the mind of man*. New York: Harper and Row.

Brewer, George R. 1970. *Ion propulsion technology and applications*. New York: Gordon and Breach Publishers.

Burgess, Eric. 1978. *To the red planet*. New York: Columbia University Press.

Bussard, R. W., and DeLauer, R. D. 1965. *Fundamentals of nuclear flight*. New York: McGraw-Hill.

Canetti, Geoffrey S. 1968. *Definition of experimental tests for a manned Mars excursion module*. Report SD-67-755-1, contract NAS9-6464. Space Division, North American Rockwell, Downey, CA.

Carr, Michael H., and Evans, Nancy. 1980. *Images of Mars: the Viking extended mission*. NASA SP-444. National Technical Information Service, Springfield, VA.

Carr, Michael H. 1982. *The surface of Mars*. New Haven, CT: Yale University Press.

Clark, Benton C. 1977. The colonization of Mars. Paper delivered at Ninth Annual Meeting, Division of Planetary Sciences of the American Astronomical Society, 27 October 1977, Boston.

————. 1978. The Viking results—The case for man on Mars, AAS-78-156. Paper delivered at 25th Anniversary Conference of the American Astronautical Society, 30 October–2 November 1978, Houston.

————. 1981. Chemistry of the martian surface: a resource for manned exploration of Mars. Paper delivered at the Case for Mars Conference, Boulder.

Clifford, Stephen M. 1981. Mars: ground ice replenishment from a sub-permafrost groundwater system. Twelfth Lunar and Planetary Science Conference Proceedings, Houston.

————, and Huguenin, Robert L. 1979. The H_2O mass balance on Mars: implications for a global sub-permafrost groundwater flow system. Tenth Lunar and Planetary Science Conference Proceedings, Houston.

Clift, Nicholas E. 1977. *A compendium of future space activities*. NASA Lyndon B. Johnson Space Center, Houston.

Cross, Douglas B., and Butts, Aubrey J. 1981. Manned Mars mission landing and departure systems. The Case for Mars Conference, 30 April–2 May 1981, Boulder.

Crouch, Donald S. 1980. Study of sample drilling techniques for Mars sample return missions. Contract NAS9-15907. Martin Marietta, Denver.

Cutts, James A. 1981. Scientific activities on the martian surface. Paper delivered at the Case for Mars Conference, Boulder.

Cutts, James A.; Blasius, Karl R.; Pang, Kevin D. K.; and Thompson, Thomas W. 1978. *Mars sample analysis mission: the role of surface mobility* (The Cutts Report). Planetary Science Institute, Pasadena, California.

Cutts, J. A.; Blasius, K. R.; Roberts, W. J.; Pang, K. D.; and Howard, A. 1980. *Analysis of polar sites*. In volume I, *Detailed reports of the Mars sample return: site selection and sample acquisition study*. Jet Propulsion Laboratory, Pasadena, CA.

David, Leonard. 1981. *The humanation of Mars*. Paper delivered at the Case for Mars Conference, Boulder.

DiPietro, Vincent, and Molenaar, Greg. 1980. *Unusual martian surface features*. Glenn Dale, MD: Mars Research.

Ehricke, Krafft A. 1968. A strategic approach to interplanetary flight. In *Fourth international symposium on bioastronautics and the exploration of space*, pp. 287–386. Brooks Air Force Base, Texas: U.S. Air Force, Aerospace Medical Division.

————. 1971. Perspectives and systems engineering of a manned planetary flight. In *Space shuttle and interplanetary missions*, AAS volume 28, pp. 337–392. San Diego: Univelt, Incorporated.

El-Baz, Farouk. 1979. Eolian features in the western desert of Egypt and some applications to Mars. *Journal of Geophysical Research*, 30 December, pp. 8205–21.

————. 1981. Desert builders knew a good thing when they saw it. *Smithsonian*, April, pp. 116–21.

Ezell, Edward C. 1979. Man on Mars: the mission that NASA did not fly. American Association for the Advancement of Science Annual Meeting, 3 January 1979, Houston.

Ezell, Edward C., and Ezell, Linda. *On Mars: exploration of the red planet*. NASA SP-4212. NASA Scientific and Technical Information Branch, Washington, DC.

Fielder, G., editor. 1975. *Volcanoes of the earth, moon, and Mars*. London: L. Wilson Publishers.

Finke, Robert C., editor. 1981. *Electric propulsion and its applications to space missions*. New York: American Institute of Aeronautics and Astronautics.

Fisher, Anna Lee, and Fisher, William F. 1980. Medical implications of space flight. TEM/Environmental Medical Emergencies, March, pp. 137–50.

Flinn, Edward A., editor. 1978. *Scientific results of the Viking project*. Washington: American Geophysical Union.

French, Bevan M. 1977. *The Moon Book*. New York: Penguin.

French, James R. 1981. Report on the results of the mission strategy workshop of the Case for Mars Conference. 26 May 1981, Boulder.

French, James R., and Burke, James D. 1981. Deep space exploration: the new challenges. *Astronautics/Aeronautics*, March, pp. 32–34.

Friedman, Louis D. 1978. *Mars sample return pre-project study*. Jet Propulsion Laboratory, Pasadena, California.

Glasstone, Samuel. 1968. *The book of Mars*. NASA SP-179. Washington, DC.

Godwin, Felix. 1960. *The exploration of the solar system*. New York: Plenum Press.

Goldman, Nathan C. 1981. The legal and political implications of the colonization of Mars. Paper delivered at the Case for Mars Conference at Boulder.

Greeley, Ronald, editor. 1979. Second International Colloquium on Mars. *Journal of Geophysical Research*, 30 December, pp. 7917–8539. Also published as NASA Conference Publication #2072, Washington, DC.

Greeley, R.; Ward, A. W.; Peterfreund, A. R.; Snyder, D. B.; and Womer, M. B. 1980. *Detailed reports of the Mars sample return: site selection and sample acquisition study*, volume II. Arsia Mons Unit, Jet Propulsion Laboratory, Pasadena, CA.

Henize, Karl G. 1979. Manned mission to Mars more feasible than it seems. *Roundup* (NASA Johnson Space Center, Houston), 23 March, p. 4.

Hollister, Walter M. 1963. Mission for manned expedition to Mars. PhD Dissertation, Massachusetts Institute of Technology, Cambridge, MA.

Hunter, Maxwell W., and Tschirgi, J. M. 1959. The advantage of high thrust space vehicles. American Rocket Society 14th Annual Meeting, 16–20 November 1959, Washington, DC.

Hunter, Maxwell W., II. 1966. *Thrust into space*. New York: Holt, Rinehart, and Winston.

Jenkins, Morris V. 1971. *Manned Mars exploration requirements and considerations*. NASA Manned Spacecraft Center, Houston.

Johnson, Richard D. 1981. The settlement of Mars as a focus for solar system exploration. Paper presented at the Case for Mars Conference, Boulder.

Kent, Stan. 1981a. Capabilities of solar electric propulsion system for a Mars exploration program. Paper presented at the Case for Mars Conference, Boulder.

———. 1981b. The Viking fund: a mandate from the people. Paper presented at the Case for Mars Conference, Boulder.

Kosmo, Joe J. 1970. *Preliminary analysis for a Mars surface space suit*. Crew Systems Division, NASA Manned Spacecraft Center, Houston.

Layton, J. Preston. 1981. *Projected space technologies, missions, and capabilities in the 2000–2020 time period*. New York: American Institute of Aeronautics and Astronautics.

Leary, Frank. 1970. Flight plan: man to Mars. *Space/Aeronautics*, February, pp. 28–38.

Lee, Vernon A., and Wilson, S. W., Jr. A survey of Mars mission trajectory characteristics. In AIAA's *Stepping stones to Mars*, pp. 35–68.

Ley, Willy, and von Braun, Wernher. 1956. *The exploration of Mars*. New York: Viking.

Lunar and Planetary Institute. 1981. *Third International Colloquium on Mars*. Held at Pasadena, California, 31 August–2 September.

McCauley, J. F.; Hipsher, H. F.; and Steinbacher, R. H. 1974. *Mars as viewed by Mariner 9*. NASA Scientific and Technical Information Office, Washington, DC.

Maguire, Bassett. 1981. Ecological problems of extended life-support on the martian surface. Paper presented at the Case for Mars Conference, Boulder.

Mandell, Humboldt C., Jr. 1981. The cost of landing man on Mars. Paper presented at the Case for Mars Conference, Boulder.

Mars to be next space goal after moon. *Aviation Week and Space Technology*, 22 July 1963, pp. 84–86.

Masursky, Hal; Dial, A. L.; Strobell, M. E.; Schaber, G. G.; and Carr, M. H. 1980. Tyrrhena Patera and Iapygia, ancient cratered terrain and candor and hebes chasmata. In volume IV, *Detailed reports of the Mars sample return: site selection and sample acquisition study*. Jet Propulsion Laboratory, Pasadena, CA.

Meyer, Tom R. 1981. Extraction of martian resources for a manned research station. *Journal of the British Interplanetary Society*. 34:285–88.

Millburn, John R. 1971. Earth/Mars orbit demonstrations. *Spaceflight*, July, pp. 259–62.

Minear, John, and Friedman, Louis. 1978. Future exploration of Mars. *Astronautics/Aeronautics*, April, pp. 18–27, 65.

Mouginis-Mark, P. J. 1980. A young-lava landing site northwest of the volcano Apollonaris Patera and a landing site on the ancient terrain southeast of the Schiaparelli Basin. In volume III, *Detailed reports of the Mars sample return: site selection and sample acquisition study*. Jet Propulsion Laboratory, Pasadena, CA.

Muson, Howard. 1980. The right stuff may be androgyny. *Psychology Today*, June, pp. 14–18.

Mutch, Thomas A.; Arvidson, Raymond E.; Head, James III; Jones, Kenneth L.; and Saunders, Stephen. 1976. *The geology of Mars*. Princeton: Princeton University Press.

Oberg, James E. 1981a. International politics of Mars. Paper presented at the Case for Mars Conference, Boulder.

————. 1981b. *Red star in orbit*. New York: Random House.

————. 1981c. *New earths*. Harrisburg, PA: Stackpole Books.

————. 1982a. Mars manifesto. *OMNI*, May, p. 13.

————. 1982b. Spaceships for Mars. *Science Digest*, May.

Oja, Heikki. 1972. Launch windows to the planets. *Spaceflight*, October, pp. 386–88.

Parkinson, Bob C. 1980. Is nuclear propulsion necessary? (or, Mars in 1995!). AIAA paper 80-1234 presented at the AIAA Joint Propulsion Conference, 30 June 1980, Hartford.

————, and Hardy, David. 1981. Mars in 1995! *Analog*, 22 June 1981, pp. 38–49.

Phinney, W.; Lofgren, G.; and Morris, R., editors. 1977. *An outline of planetary geoscience*. NASA TMX-58202, Lunar and Planetary Sciences Division, NASA Johnson Space Center, Houston, Texas.

Quattrone, Philip D. 1981. Extended mission life support systems. Based on Boulder speech at Ames Research Center, California.

Singer, Fred. 1981. The Ph-D proposal: a manned mission to Phobos and Deimos. Paper presented at the Case for Mars Conference, Boulder.

Snyder, Conway W. 1979. The planet Mars at the end of the Viking mission. *Journal of Geophysical Research*, 30 December, pp. 8487–519.

Staehle, Robert L. 1981. An expedition to Mars employing shuttle-era systems, solar sails, and aerocapture. Paper presented at the Case for Mars Conference, Boulder.

Stevens, Charles B. 1980. Using fusion for propulsion. *Fusion*, October, pp. 30–32.

Stevenson, John B. 1980. Concept for establishing four-person Mars bases in this century. AIAA Houston Section Fifth Annual Technical Mini-Symposium, Houston.

Stine, G. Harry. 1982. Onward to Mars! *Analog*, 1 February, pp. 60–71.

Vajk, J. Peter. 1979. *Planetary exploration space colony style*. Science Applications, Inc. Pleasonton, CA.

Veverka, Joseph, and Burns, Joseph A. 1981. The moons of Mars. In *Annual review of Earth and planetary science 1980*, pp. 527–558.

Washburn, Mark. 1977. *Mars at last!* New York: Putnam.

Wells, Ronald Allen. 1979. *Geophysics of Mars*. New York: Elsevier Scientific.

Woodard, Daniel, and Oberg, Alcestis. 1981. Medical aspects of a flight to Mars. Paper presented at the Case for Mars Conference, Boulder.

Zisk, S. H., and Mouginis-Mark, P. J. 1981. Oasis revisited: further analysis of the Solis Lacus radar anomaly on Mars. In *Abstracts of the Twelfth Lunar and Planetary Science Conference*, Lunar and Planetary Institute, Houston.

Index